Praise for *DESERT WISDOM/AGAVES* and *CACTI*

"This is a great book to learn about the uses and physiology of cacti and agaves. It describes the main features of these extraordinary desert plants using an easy, clear, and humorous writing style."

—**José C. B. Dubeux Jr.,**
Universidade Federal Rural de Pernambuco,
Recife, Brazil

"Considering the world's problems in arid regions, this contemporary explanation of the benefits of cacti and agaves by the world's authority on the ecophysiology of these plants is most welcome and needed."

—**Peter Felker,**
D'Arrigo Bros. Co,
Salinas, California, U.S.A.

"Park Nobel is back again on the shelves, sharing his love of desert plants! Fasten your seat belts and fly over the ages and continents with fantastic stories and the unique world of Agaves and Cacti."

—**Paolo Inglese,**
Università degli Studi di Palermo,
Palermo, Italy

"Equally important to the curious, to academics, and to decision makers, this crystal clear book combines updated scientific information on global warming with possible solutions that involve these incredible Crassulacean Acid Metabolism (CAM) plants."

—**Yosef Mizrahi,**
Ben-Gurion University of the Negev,
Beer Sheva, Israel

"Park Nobel is the international 'guru' for the ecophysiology of cacti and agaves. This new book adds to his growing list of outstanding research publications and special books on the topic."

—**Ali Nefzaoui,**
ARDA North Africa Program,
Tunis, Tunisia

"With graduate students and scientists from around the world, Park Nobel has pioneered studies on CAM plants in arid regions. Their carbon gain and water relations will allow them to thrive in environments brought about by global climate change, as cogently described in this new book."

—Eulogio Pimienta-Barrios,
Universidad de Guadalajara,
Guadalajara, Jalisco, Mexico

"Drawing on previous more technical efforts, Park Nobel, the world's foremost authority on the functional biology of cacti and succulents, has yet another book on these fascinating plants. It synthesizes key information, including climate change effects, in a lucid style comprehensible by the general public. A must read!"

—Stanley D. Smith,
University of Nevada,
Las Vegas, Nevada, U.S.A.

DESERT WISDOM/ AGAVES and CACTI: CO$_2$, Water, Climate Change

PARK S. NOBEL

Distinguished Professor of Biology Emeritus
Department of Ecology and Evolutionary Biology
University of California, Los Angeles, California
psnobel@biology.ucla.edu

iUniverse, Inc.
New York   Bloomington

iUniverse books may be ordered through booksellers or by contacting:

iUniverse
1663 Liberty Drive
Bloomington, IN 47403
www.iuniverse.com
1-800-Authors (1-800-288-4677)

Because of the dynamic nature of the Internet, any Web addresses or links contained in this book may have changed since publication and may no longer be valid. The views expressed in this work are solely those of the author and do not necessarily reflect the views of the publisher, and the publisher hereby disclaims any responsibility for them.

ISBN: 978-1-4401-9151-0 (sc)
ISBN: 978-1-4401-9152-7 (ebook)

Library of Congress Control Number: 2009912502

Printed in the United States of America

iUniverse rev. date:12/09/09

CONTENTS

PREFACE

Why consider agaves and cacti? To answer this question, this brief primer relates their survival, net CO_2 uptake, and biomass productivity, often under harsh conditions, to some of the greatest environmental challenges of our era—the upcoming increases in air temperature and the changes in rainfall patterns. Such production of biomass has many uses, ranging from cattle feed to industrial chemicals to biofuels to preventing desertification to a sink for global carbon dioxide. The scientific literature often does not satisfy the quest for accurate, comprehensible, readily available information. Yet the extraordinary biomass productivity of agaves such as *Agave tequilana* (known for tequila, the key ingredient of Margaritas) and cacti such as *Opuntia ficus-indica* (the "Indian fig," known for its delicious fruits) under relatively dry conditions should interest many and will surprise some.

Why consider agaves and cacti together? You may know that these two groups of plants are unrelated taxonomically. Indeed, agaves are more closely related to grasses and cacti are more closely related to apple trees and roses than they are to each other. Nevertheless, some people confuse agaves and call them "cacti" because like cacti they have spines, are succulent, and occur in deserts. Agaves are the largest genus in the monocot family Agavaceae. And 'cacti' is a synonym for its dicot family, the Cactaceae. We will consider them together because they are often similar physiologically and ecologically. Physiologically they each use the same photosynthetic pathway out of the three that are possible. Ecologically they can survive with limited amounts of water and often in very hot environments—it's their efficient use of water that is of utmost importance.

A theme throughout this book is global climate change, which has been triggered by the clearly rising atmospheric CO_2 levels. We also discuss Crassulacean Acid Metabolism (CAM), the intriguing photosynthetic pathway used by agaves and cacti whose key feature is the nocturnal opening of stomata and the nocturnal uptake of CO_2. CAM leads to less water lost by transpiration. This is embodied in a benefit/cost index known as the "Water-Use Efficiency," which is the CO_2 fixed via photosynthesis divided by the water lost via transpiration. An "Environmental Productivity Index," which was proposed and developed by me, quantifies how light, temperature, soil water, and nutrients affect CO_2 uptake by plants. Various agaves and cacti can produce extremely well under limited water availability, low soil fertility, and often high temperatures. These conditions, which can nearly eliminate biomass productivity by most plants, will increase in the future as the global climate changes.

So why should you read this book? Perhaps you already have an appreciation of the beautiful forms and the many uses of agaves and cacti—if not, you are in for a treat. Plant names, current cultivated areas, and productivity (in both international and U.S. units) are indicated. Your questions about global climate change are anticipated, and answers are given about the expected responses of agaves and cacti. For some readers the key "take-home" lesson will be how to predict the effects of known and projected environmental changes on their success and potential productivity in specific regions of interest to them. With humor and insight, a basic approach to the physiology of these CAM plants plus predictions about future climates should be of interest to all in light of the resiliency of agaves and cacti.

My 1994 book *Remarkable Agaves and Cacti* (Oxford University Press, New York; translated into Spanish as *Los Incomparables Agaves y Cactos* by Edmundo Garcia Moya, 1998, Trillas, Mexico City) had only one page devoted to "global climate change." Now this topic is part of the title, most of two chapters, and a recurrent theme. The immediate stimulus to write the present book came from *Perspectives in Biophysical Plant Ecophysiology: A Tribute to Park S. Nobel* edited by Erick de la Barrera and William K. Smith (2009, Universidad Nacional Autónoma de México, Mexico City). Its Epilogue challenged me to assess the future of agaves and cacti with respect to climate change. Another stimulus was the numerous questions over the years from friends and the media on the possible survival and the productivity of such CAM plants under stressful situations, which had provocative but often exaggerated claims in the literature.

For comments and input that greatly improved this often personal book, special thanks are due to Edward Bobich, Philippa Drennan, Carol Felixson, Edmundo Garcia Moya, Gretchen North, Barry Osmond, Lawren Sack, Andrew Smith, Jennifer Sun, Brian Zutta, and especially Catherine Goodman. It was a joy and challenge to write, and hopefully you will enjoy and learn from reading it.

Park S. Nobel
November 4, 2009

1

Current Uses of Agaves and Cacti

To hint at the multiple possibilities for the future uses of agaves and cacti that take into account climate change, we begin with their current uses. These range from food (for humans) to fiber to forage (where the animals search in the field) to fodder (where the material is brought to the animals) to fruits to fructose (for diabetics) to funny stuff (e.g., dyes, hormones, and hallucinogens) to fancy plants (ornamentals) to friendly beverages (from aguamiel to tequila). We start with the genus *Agave*, containing about 140 species. *Agave* is the largest genus in family Agavaceae, whose name is based on the Greek for "noble" (*agauos*) to recognize the group's many ancient uses by humans. We then tackle the much larger family Cactaceae, containing about 1,600 species. Cactus (family Cactaceae) comes

1

from the Greek *kaktos*, referring to a spiny, thistle-like plant. These two taxa contain many remarkable plants.

Figure 1-1. Various agaves and cacti growing in the author's yard in Los Angeles, California. Note the cladodes (flattened stem segments) on the platyopuntias in the foreground and the leaves radiating from the base of the agaves in various locations. Planting began 15 years earlier on sloping ground with relatively infertile rocky soil. The local median annual rainfall in this semi-arid area was 380 millimeters (15 inches) for the period involved.

Much of the optimism for the future of agaves and cacti rests with *platyopuntias*, a group of cacti in the genus *Opuntia* that has flattened stem segments referred to as *cladodes* (see Fig. 1-1). Beginning long ago, cladodes were detached from plants in the wild and placed in the ground near houses. Admirably, such planted hunks grew into tall plants in a few years. Anyone could have noticed their potential for fodder and food, as rodents, rabbits, wild pigs, and deer ate them voraciously despite the spines. As an added benefit, platyopuntias (or "opuntias," for short) produce edible fruit.

Another resource appreciated long ago was the large leaves of agaves (Fig. 1-1), which were harvested for thatch roofs and for their fiber, which was used for clothing, bindings, and ropes. The potential for beverages was probably also discovered at an early time, perhaps by accident. A sweet fluid could have exuded from a cut or damaged agave inflorescence (flowering stalk), which soon fermented on its own. Among agaves and cacti, opuntias grown for fodder currently occupy the most land area and offer great promise for the future. Other uses are more entertaining, as we shall see as we discuss the main current uses of agaves and cacti.

Agaves

Aguamiel and Pulque—Sweet to Fermented

My first encounter with the non-alcoholic liquid *aguamiel* (Spanish for *honey water*) occurred in the 1980s. In the early summer, kids in various parts of Mexico were selling a beverage in roadside stands. Curious, I sampled this rather sweet drink with a heavy flavor of agave leaves. Anyone with a sweet tooth would like it. But many prefer what happens after aguamiel is allowed to ferment for a few days, either naturally, as in the past, or with the injection of a small amount of a current batch of the alcoholic beverage *pulque*. Aguamiel is also incorporated into one of the passions of our times, energy drinks.

Pulque and aguamiel are produced from sap collected from certain agaves, especially near Mexico City, a tradition begun 2,000 years ago. The agaves include *Agave atrovirens*, *Agave mapisaga*, *Agave salmiana*, and the ubiquitous *Agave americana*, as well as about eight other species. *Maguey* is commonly used to refer to various such species in Mexico; maguey is a term of indigenous origin traced to the first agaves seen by Spaniards in the Caribbean in the 15th and 16th centuries.

Let me take a small diversion to comment on plant names. In the Latin binomial nomenclature, the first word is the genus and the second word is the species epithet, both of which by convention are italicized. The genus must start with a capital letter, as in *Agave tequilana* or *Homo sapiens* (us). When a Latin binomial occurs a second time, the genus can be abbreviated to its first letter (except at the beginning of a sentence), a convention that we will follow within a section or a subsection. For those who are concerned about the authorities for scientific names, the Latin binomials used in this book are based on Gentry (1982) for agaves and Anderson (2001) for cacti.

Getting back to the story, let us indicate how aguamiel is collected and pulque is produced, processes known to the ancient Aztecs (Nobel, 1994). In the 13th century, they had a goddess of pulque, Mayahuel, depicted in colorful frescoes. And pulque was used in their religious ceremonies. To collect the liquid needed for aguamiel, the incipient inflorescence of an agave a dozen years old or older is removed. (Agaves are perennial plants but essentially all of them flower only once; that is, they are *monocarpic*, as they die after their single flowering.) Flowering is usually a very showy event—the inflorescence can be 1.5 to 10 meters (5 to 33 feet) tall. Such a large size requires a tremendous amount of resources, meaning that the "goodies" that could have been used to produce beverages are commandeered for the enormous reproductive effort. For beverage production the inflorescence is hence cut off, a process called *castration* (not entirely appropriate, but it gets the point across).

In any case, after "castration," a hemispherical basin is carved in the center of the agave plant. No longer needing to supply the inflorescence, the leaves then direct all of their photosynthetic effort to the basin. The resulting liquid that exudes into this basin is collected, often using hollow gourds, a couple of times a day for many months. A single plant can produce 700 liters (185 U.S. gallons) of aguamiel. Wooden casks containing the aguamiel are carried by burros and other pack animals and then emptied into a vat, where *fermentation* proceeds, usually for a few weeks. (Fermentation refers to the chemical conversion of carbohydrates into alcohols, generally in the absence of oxygen and in the presence of yeast. *Carbohydrates* are compounds containing carbon, hydrogen, and oxygen, such as the products of photosynthesis.) About 20,000 hectares (50,000 acres) are devoted to aguamiel/pulque production in Mexico, leading to about 200 million liters (55 million U.S. gallons) of these beverage annually.

Pulque is sold as well as consumed in "pulquerias." To broaden its appeal, it is often flavored with bananas, cherries, mangoes, strawberries, or pineapple. These additions can also spruce up its color from an otherwise dullish white fluid. I will never forget the young girl of about six bravely entering a pulqueria to get her plastic container filled for her grandfather, as this beverage is especially appreciated by an older generation. Also, I have met many indigenous Mexicans who swear by pulque. One gentleman told me that the amino acids, sugars, and vitamins in pulque can sustain him for weeks and its alcohol kept him happy. Some of the clergy in Mexico in the 15th century were heavy consumers of pulque. With its relatively low alcohol content of 3 to 4%, such a beverage would allow them to go about their blessings in a blissful state. Before the advent of purified water in Europe around the Middle Ages, many people there also got most of their liquids from low-alcohol fermented beverages, often produced from honey or fruit (the alcohol supposedly reduced the bacteria rampant in water at that time).

Mescal and Tequila—Elixir of the Gods

Yes, we might all like to walk around in a happy state. But what if we were to release our inhibitions by drinking a beverage with an even higher alcohol content than pulque? It could be mescal or tequila. But not so long ago the dangers of alcohol abuse led to the Prohibition era (1920 to 1933) in the United States, when it was not legal to sell alcoholic beverages. Prohibition was known as "The Noble Experiment" in the literature (it was not called "The Nobel Experiment").

Prohibition was institutionalized by the 18th amendment to the United States constitution, passed at the end of 1919. It was repealed by the 21st amendment passed in 1933, with taxation and various restrictions thereafter being used for control of alcoholic beverages. Periods of prohibition have also occurred in many northern European countries in the 20th century. Moreover, strict Islamic Law forbids the consumption of alcoholic beverages, although the strictness of the enforcement of this prohibition varies greatly among Muslim nations. Let us now dive into mescal (also spelled *mezcal* in English, which is actually the Spanish spelling) and tequila, which are now popular alcoholic beverages produced from agaves.

To produce mescal and tequila, we begin with essentially the same source of starting material as for pulque, namely, carbohydrates in the stems of agaves. To increase the average alcohol content of

3 to 4% characteristic of pulque to the average alcohol content in mescal (40 to 50%, 80 to 100 proof) or tequila (35 to 45%, 70 to 90 proof) requires *distillation*. Distillation means heating a fermented brew to vaporize various compounds. As the vapor is cooled, certain compounds condense out at different temperatures and pressures, such as water and "alcohol," which in this context means ethanol (CH_2CHOH). Many other alcohols also occur in the vapor, such as methanol (CH_3OH), which can be unpleasant in small doses and deadly in high doses. Also, many heavier alcohols condense out, as do various other organic molecules, giving each brew a different flavor, as is well known for the many types of whisky. In this regard, mescal is traditionally singly (once) distilled but increasingly doubly (twice) distilled; tequila is doubly distilled (Gentry, 1982; Valenzuela-Zapata and Nabhan, 2004). Double distillation means that the original distillate is heated and vaporized again, as is also the case for singly versus doubly distilled whiskies.

Mescal is produced from about a dozen species of agaves, especially *Agave angustifolia*. Although large-scale commercial production is increasing, a person can still bring his or her own bottle to picturesque stills owned by rural families in the outskirts of the city of Oaxaca in the southern Mexican state of Oaxaca. Mescal is also produced in the northwestern state of Sonora, using *A. angustifolia* and *Agave palmeri*, while *Agave salmiana* is a popular source of mescal in central Mexico. Perhaps 30,000 hectares (75,000 acres), often intercropped with beans or corn (maize), are used to produce about 20 million liters (5 million U.S. gallons) of mescal annually. Bottles of certain mescals often famously (should we say "notoriously"?) contain the larva of two types of insects that feed on agaves. Such "worms" enhance the marketing uniqueness and hence appeal of this beverage.

The main locus for tequila production is in the western state of Jalisco, Mexico. In Jalisco and neighboring states, about 80,000 hectares (200,000 acres; Nobel, 2003) of *Agave tequilana* are cultivated. The favored variety is known as *azul* (*blue* in Spanish), recognizing the attractive bluish tint of its leaves. Tequila production began near a city in Jalisco named Tequila after the introduction of distillation by the Spanish in the 16th century. The beverage tequila may be the first distilled spirit indigenously produced in the New World. In any case, to protect the industry, it is now illegal to export plants of *A. tequilana* from Mexico for commercial purposes.

A beverage can be called "tequila" only if over half of its alcohol comes from sugars derived from *A. tequilana* and less than half come from cheaper sources of fermentable sugars (or ethanol added directly), such as from sugar cane or corn (maize). *Agave tequilana* is often cultivated in hilly regions where irrigation is not feasible. Indeed, it grows well in Jalisco in regions with 450 to 800 millimeters (18 to 31 inches) of rainfall annually but no nighttime freezing temperatures. We will return to the environmental aspects for the cultivation of agaves in later chapters.

Margaritas are a drink enjoyed worldwide containing tequila, lime or lemon juice, crushed ice, and usually orange liqueur (Triple Sec). According to one among many stories, it was invented in the 1940s in Tijuana, Baja California Norte, Mexico, to honor an American showgirl named "Margarite." After blending the ingredients, a Margarita is usually served in a wide-brimmed glass whose moistened rim has been dipped in salt.

Tequila is actually consumed in many ways and in many different drinks. Brands based only on agave and carefully aged for many years can cost $100 (U.S. dollars) per bottle. They are often consumed "neat" (nothing added), as is also the case for expensive brandies and cognacs. After a long day's work measuring the size and the weight of leaves of *A. tequilana* in the field and desiring a beverage a little less potent than tequila neat, I went to a restaurant in Tequila and ordered a glass of "vino blanco" (*white wine*). The waiter looked at me a bit aghast and promptly brought me a large glass of a young tequila.

To harvest *A. tequilana* for tequila production, a mature plant usually 5 to 8 years old has its leaves cut off near the base of an enlarged stem just before the inflorescence appears and flowering occurs. A similar technique is used for the somewhat older agave plants used for mescal production, such as *A. angustifolia* and *A. salmiana*. These procedures lead to a form not unlike the shape of a pineapple, hence the name "piña" (*pineapple* in Spanish) for the harvestable part.

A piña generally weighs 20 to 60 kilograms (45 to 135 pounds). It is chopped into small pieces, and is then roasted for 24 to 48 hours. Roasting breaks down the starch and other similar polymers into their small carbohydrate subunits, such as glucose and fructose. These can be directly used for "agave syrup" (also called "agave nectar"), whose non-water content is often 75% fructose. After fermenting the carbohydrates for 1 to 3 days, the alcohol is concentrated by distillation. The more expensive tequilas are then aged for a few months to a few years in oak barrels. In any case, about 70% of the annual production in Mexico of just over 300 million liters (80 million U.S. gallons) is exported. Thus, if you wish, you can order a Margarita in restaurants and bars around the world.

Fiber—Great Before Synthetics Arrived

The fibrous nature of agave leaves is obvious to anyone who has touched these plants—the stiff leaves resist pushes and shoves. Besides generally having spines along the leaf edges, an agave leaf also terminates in a sharp point or spike. These tips have been used as needles in the past and readily penetrate the flesh of those who would carelessly study agaves in the field today. Indeed, agaves have been planted in various countries as barriers to hinder the intrusion of robbers into yards and the advance of military invaders. Such leaves can be huge. For instance, leaves of *Agave mapisaga* and *Agave salmiana* can weigh 40 kilograms (90 pounds) each, which is not your ordinary ivy leaf.

Agaves are monocots, a type of angiosperm (flowering seed plant with fruits) having one seed leaf per embryo and parallel veins in the leaves. When mature, these veins in the leaves are the fibers of commerce. The veins are actually vascular bundles that carry water from the soil (in the xylem) and distribute photosynthetic products around the plant (in the phloem). Their strength comes from the cell walls surrounding each plant cell. Such cell walls consist primarily of the glucose polymer cellulose (more about its digestibility in Chapter 7). The cell wall can also be impregnated with lignin, a complex polymer that together with cellulose are the principal components of wood; lignin resists degradation and so helps create coal over geologic time periods. Although lignin is involved in the stiffness of cells, cellulose actually gives the strength to the cells. Indeed, the cellulose polymers greatly resist being stretched. Consider a cotton shirt, which is almost pure cellulose. You could say to someone, "Hey, you look really great in that cellulose!" But be careful, as nowadays many shirts and other articles of clothing also contain or are made entirely of synthetic fibers.

The demise of the commercial use of fibers from agaves coincided with the advent of synthetic fibers. Agave fibers were not used for shirts, but instead were used for bindings, nets, sacks, twines, ropes, upholstery padding, carpet pads, mats, baskets—the list goes on and on (Nobel, 1994). The Achilles' heel of agave fibers was and is their weakening upon being wet, while synthetic fibers retain their strength in the presence of moisture. Moreover, synthetic fibers are uniform, often stronger, and also not subject to rotting when wet. Synthetic fibers are usually produced by extruding a liquid under pressure through fine holes, leading to threads, as for nylon, which became available in 1938. Other common synthetic fibers developed more recently include acrylics, polyesters, Spandex, and numerous other trade-marked threads.

As just alluded to, the agave fiber industry was once big business worldwide. In Mexico, the species of choice to cultivate was and is *Agave fourcroydes* (common name *henequen*). And the region of choice was and is the Yucatán, a large state on its eastern coast. Yet the most widely used species for fiber in Mexico is *Agave lechuguilla* (common name *lechuguilla*), which is harvested in nine Mexican states, nearly all from the wild. Actually, *A. lechuguilla* has more plants in the wild than does any other species of agave.

Outside of Mexico, the favored species for fiber production is *Agave sisalana* (common name *sisal*). It has been cultivated in the Caribbean, Brazil, India, many Pacific islands, Australia, and parts of Africa. Sisal was first exported from the port of Sisal in Yucatán, Mexico. At the end of the 19th century, the German East Africa Company was responsible for plants going from Sisal to Florida to Germany to German East Africa (now Tanzania). The 62 sisal plants that survived these perilous journeys became the basis for the commercial cultivation of *A. sisalana* in seven countries in southern and especially eastern Africa. In the early 20th century, about 70% of the world's long, hard, plant fibers came from *A. sisalana*.

Both *A. fourcroydes* and *A. sisalana* produce about 200 leaves before sending up an inflorescence, their last gasp before dying. This occurs at about 8 to 14 years of age. Before this, the leaves, which can be 1 meter (39 inches) long, are removed and sent to factories to remove the fiber. Either by crushing, scraping, or rasping, the soft pulp is eliminated. The leaf juices can also be collected at this stage, which can be the starting material for certain industrial chemicals to be considered shortly. The remaining fibers are then dried, often in the sun for these two species and for *A. lechuguilla* as well, which also bleaches any remaining chlorophyll.

Agave cultivation has dropped dramatically with the advent of synthetic fibers, especially just after World War II. Perhaps 100,000 hectares (250,000 acres) worldwide still have some commercial production of agave fibers, down from over 1 million hectares (2.5 million acres) in its heyday. Yet certain attractive products from agave fibers still persist, such as woven mats, sandals, and handbags, made mostly in cottage industries.

Fodder—An Underexploited Resource

Often a technology needs the use of byproducts to be viable. A case in point is the production of fibers from agave leaves. Perhaps the soft pulp removed from the leaves, as just discussed, could be fed to cattle. This is actually not a "perhaps," as I have witnessed this successful application in Mexico and Tanzania. Also, I have seen leaves of agaves carried by trucks in Mexico from fields to feed both goats and cattle.

Relatively little scientific literature apparently exists on the topic of agave leaves as fodder. It certainly merits attention, especially in light of our discussion of regions where agaves and cacti can be cultivated in the future (Chapter 7). In this regard, the exploitation of opuntias for both forage and fodder is well developed, as we will discuss shortly. So this discussion is purposefully brief, as I think that the best is yet to come for agave leaves as fodder.

Industrial Chemicals, Birth Control Pills

We have hinted that the juice squeezed out of leaves of *Agave fourcroydes*, *Agave lechuguilla*, and *Agave sisalana* during fiber preparation might contain something chemically interesting. Indeed, agave leaves contain a high content of sapogenins, which are steroids and closely related triterpenes (moderately complex organic molecules containing 17 to 30 carbon atoms). Steroids, which can act as hormones, can be synthesized through controlled chemical processes. But wouldn't it be nice to have them produced much more cheaply by a plant? Actually, only a few chemical steps are necessary to convert agave sapogenins into steroids such as the anti-inflammatory cortisone or the hormones estrogen and progesterone. Indeed, many plant compounds will probably play increasingly important roles in future medicine, as the wisdom gained over the ages moves from local anecdotes to wider acceptance to Phase III clinical trials to your doctor's next prescription. But let us first take a few steps back into recent history.

Estrogen plays many important roles in human physiology, especially at various life stages for women. When its level and that of progesterone become high, ovulation is inhibited, so it can be a potent birth-control pill. Before organic chemistry made such great strides in synthesizing molecules after World War II, estrogen was not readily available at a reasonable price. Yet *Agave americana*, *Agave vilmoriniana*, and some other agaves can contain 2 to 5% sapogenins per unit dry biomass, and these sapogenins can be fairly easily converted to estrogen and progesterone. This was exploited on a limited scale in the late 1940s for oral contraceptive purposes in the United States.

Many other such compounds in agave leaves, stems, and roots may be converted to useful hormones in the future. Undoubtedly other interesting chemical substances also exist in the 140 species of agaves, which could be produced cheaply and naturally. For instance, some essentially naturally occurring steroids might lead to weight gain (muscle development) in poultry. They could similarly stimulate weight gain by goats, sheep, and cattle but perhaps with low levels of cholesterol and poly-unsaturated fats. These speculations are meant to be challenges to people who deal with animal feed and what we may expect for improved meat production in the future using presently unexploited compounds in agaves.

Ornamentals—Leaves, Inflorescences

This topic really requires few words—it is more personal and visual (Fig. 1-1). Indeed, agaves are planted and admired in more than 100 countries worldwide and on all continents (except Antarctica) for the beauty of their leaves radiating from the base and the tall showy inflorescences. They are grown mainly outdoors, even where mildly freezing temperatures can occur.

Many species have already been mentioned, such as *Agave americana*, which is often simply called a *maguey* and also is known as the *century plant* because of its relatively long life span before flowering. It has varieties with striking white or yellow stripes along its leaves, namely, *marginata*, *medio-picta*, *striata*, and *variegata*. The species was introduced from Mexico into Europe in the 16th century; there it is often moved indoors during the winter. The striped forms of *A. americana* are perhaps the most treasured agaves as ornamentals worldwide.

Many other agaves are also used as ornamentals. *Agave attenuata* (Fig. 1-1) is extensively admired in the southwestern United States and other regions, because it grows readily, produces numerous offshoots, and has relatively flexible leaves with no spines along their margin and a soft leaf tip. Hence it is a welcome addition along pathways—it does not deter invaders but rather devotes its energy to beauty, including an arching inflorescence about 3 m (10 feet) tall (Fig. 1-1). The many flowers along its inflorescence also attract bees and hummingbirds.

Small agaves that make delightful additions to gardens include *Agave filifera*, which has attractive fibers projecting from the margins of its leaves. Leaves of *Agave potatorum* are dramatically spatula shaped, as are those of *Agave parryi (Agave huachucensis)* and *Agave pumila*. *Agave victoriae-reginae* (Queen Victoria's agave) is another attractive small agave that has spectacular whitish lines along its leaves. Increased use of ornamental agaves in the future, especially in gardens where their requirements for only small amounts of water is an obvious advantage, should parallel the increase in the use of the highly productive ones for other reasons (Chapter 7).

Cacti

Fruits—Colorful and Delicious

Cactus fruits have been enjoyed since prehistoric times in present-day Mexico. Now opuntias, primarily *Opuntia ficus-indica* (the "Indian fig"), are commercially raised for their fruit in over 20 countries worldwide. Formerly the plants and their fruits were known as *prickly pears.* But since about 1990 the name has been shifting to *cactus pears* to avoid calling attention to the spines that are present on the fruits as well as on the cladodes.

Currently about 70,000 hectares (170,000 acres) are devoted to cactus pear in Mexico, the country with over half of such world's fruit production. By far the main species is *O. ficus-indica,* whose fruits have various ripe internal flesh or pulp colors, with yellow-orange being the most common and green being especially delicious. With minimal management, production can annually average 5 to 8 tonnes of fruit per hectare (2.2 to 3.6 tons per acre; 1 tonne, sometimes called a metric tonne and even a metric ton, is 1,000 kilograms or 2,205 pounds; 1 ton is 2,000 pounds).

In Italy, more specifically Sicily, about 18,000 hectares (45,000 acres) of *O. ficus-indica* are cultivated for fruit. The ripe pulp colors are yellow-orange (the most common commercial one), red-purple, or white-cream. Fruit production in Sicily, which is the most important cactus pear area in the Mediterranean region and the second most important in the world, can annually yield 20 to 30 tonnes of fruits per hectare (9 to 13 tons per acre) under intensive plantation management (Inglese, Basile, and Schirra 2002).

Consumption of fresh cactus fruits in Mexico and Sicily averages 3 to 4 kilograms (7 to 9 pounds) per person per year (other uses of such fruits will be discussed shortly). Substantial production of cactus pears also occurs in South Africa, Tunisia, Chile, and Argentina. Opuntias are raised for fruits on a more limited basis in California and other parts of the United States, where red-purple pulp colors are greatly appreciated. In any case, fruits can represent about 40% of the above-ground biomass productivity of the plants, similar to other fruit crops (Inglese et al., 2002).

Two stumbling blocks occur for the ubiquitous acceptance of such cactus fruits. The first is the spines and especially the nasty glochids, which are very fine spines that can readily stick into your skin when you touch a freshly picked fruit. These can removed by abrasion (brushing, usually now commercially done by machines). The second stumbling block for universal acceptance of cactus pears is the seeds, which are approximately the size of grape seeds and are similarly hard.

My first introduction to these delicious cactus fruits and their seeds came in Santiago, Chile, in the early 1980s. My host had removed the rind from chilled fruit by a few deft knife strokes. The resulting pulp, about the size of a large plum or an apricot, had been quartered and placed on a plate. His two daughters and I dug in. They simply picked up a quarter, plopped it in their mouths, chewed a little, and then swallowed. I laboriously tried to pick out the seeds. The two kids had finished a fruit each while I was still struggling with my first quarter. To compete successfully, I had to swallow the seeds, which I now do without hesitation. The seeds simply pass on through, probably helping the digestive tract in the process.

The fruits are produced around the edge of a cladode. Three to over twenty flowers that subsequently lead to fruits can develop on a single, one-year-old cladode in the spring. Healthy, well-positioned cladodes usually produce larger fruits, which bring a higher market price. The fruit load per cladode can also be thinned by removing excess fruits, leaving no more than ten fruits per cladode. Such fruits become larger than they would have been before some of their competitors were removed. Hence they are commercially more valuable.

More drastically, all of the fruits can be removed in the spring, which causes the plant to flower again and to lead to a much later harvest, usually in the late autumn/early winter. At this time of the year, there is little competition from other fruits, so a much higher price can then be realized for the cactus pears. This procedure has a very interesting origin. According to one 19th century story, a rather hot-blooded Italian gentleman was so mad at the advances of his neighbor's son toward his daughter that he knocked off all of the fruits on his neighbor's opuntias. Although the neighbor was dismayed at first, the second flush of fruits proved to be much more valuable. Thus one man's rage led to a technique that is now institutionalized as the main way of producing commercial cactus pears in Sicily. It currently is also used in many other locations worldwide.

Opuntia ficus-indica is cultivated for cactus pears mainly in regions with mild winters and limited annual rainfall. In arid regions, where the annual rainfall is less than 250 millimeters (10 inches), irrigation is necessary for its economic cropping, especially if no rainfall occurs just prior to and during fruit development. In Mexico, cactus pear production is feasible in semi-arid regions (250 to 450 millimeters, 10 to 18 inches of annual rainfall) without irrigation, as sufficient rainfall occurs there during the crucial fruit development period. Plantations are usually established using detached mature cladodes with one-third of their surface area placed below ground. Although high summertime temperatures are not important for choosing such sites, the sites should be regions without substantial periods of wintertime freezing temperatures.

Besides cactus pears, many other cactus fruits are regularly eaten. Two such categories have fruits that can be found in markets in about 20 countries. The first category involves medium to tall columnar cacti, whose fruits are termed *pitayas*. The second and more important category with regard to commercial fruits embraces epiphytic and hemiepiphytic vine cacti, whose fruits are termed *pitahayas*. [*Epiphytes* are plants that grow on other plants, such as in tree crotches; *hemiepiphytes* (sometimes spelled *hemi-epiphytes*) can grow on other plants but also can extend their roots into the soil.] Considering pitayas first, these cactus fruits can be found to a limited extent in markets in Latin America, although many more are collected from the wild for personal use. In Mexico, various species of *Stenocereus* are cultivated, especially *Stenocereus queretaroensis*. In Israel and the United States, *Cereus peruvianus* (*Cereus repandus*) is cultivated to a small extent for pitayas (Nerd, Tel-Zur, and Mizrahi, 2002). The fruit peel of pitayas can be green or red and the pulp can be white or colored, often an attractive reddish.

The most commonly cultivated pitahaya is *Hylocereus undatus*, whose fruits occur in markets in about a dozen countries. Also cultivated are other species of *Hylocereus* and *Selenicereus megalanthus*. In 1860 *H. undatus* was introduced by the French to Vietnam, where it is now a main crop in the Mekong Delta region near Ho Chi Minh City (formerly called Saigon). It is often known as "Dragon" fruit or "Red Dragon" fruit and has many delightful features. First, it has no glochids and few or no spines. Second, its seeds are readily ingested, as they are the size and consistency of kiwi seeds. Third, the fruit is large, often 300 to 500 grams (10 to 18 ounces). Fourth, the fruit is beautiful—red outside (peel) and white inside (pulp). Other pitahayas can have a red peel and a red to deep-red (even violet) pulp, a yellow peel and a white pulp, or a green peel and various colors of pulp. They are truly tasty and colorful.

To me pitahayas tasted delicious the first time I tried them. This was in the 1980s in Colombia, where they are native as well as to other tropical regions of Latin America. Another memorable tasting of *H. undatus* occurred in a café in Ho Chi Minh City, where I was watching an intense televised international football (soccer) match between Vietnam and its arch rival, Cambodia. When the enthusiastic viewers got a little too rowdy, I produced a few Red Dragon fruits. The waiter cut them up and passed them around. Everyone soon calmed down and showed surprise that a foreigner knew about such delicious fruits.

Hylocereus undatus can be grown unshaded outside (e.g., in Vietnam) or in partial shade on racks under trees (e.g., in Colombia). Because it does not fare well under high sunlight, in Israel it is grown under shade nets that reduce the incoming light by 20 to 60%. It begins to show damage at chilling temperatures below 5°C (41°F) and is especially vulnerable to freezing temperatures. With drip irrigation and close spacing on trellises, annual yields can be 40 tonnes of fruit per hectare (18 tons per acre). The cultivation of pitahayas has surpassed 20,000 hectares (50,000 acres) worldwide, with substantial areas in Vietnam, Israel, Nicaragua, Guatemala, Mexico, Colombia, and Ecuador. It is a very promising fruit crop for the future, and the area of cultivation is expanding rapidly into other countries in South America as well as in Thailand and in Australia.

Nopalitos—Vegetables with Fructose
While doing research in Mexico on agaves and cacti over the years, I have had many opportunities to eat the tender young cladodes of platyopuntias, *nopalitos*. These nopalitos were mainly from *Opuntia ficus-indica*, but also from *Opuntia robusta* plus other

species of *Opuntia* and the closely related genus *Nopalea*. Indeed, at one scientific meeting in the 1990s, a highlight was the 80 different dishes prepared from nopalitos, ranging from salads to soups to main dishes to desserts. Yet this delicious culinary habit has not caught on worldwide, although several supermarkets in the southwestern United States and a few in Europe sell the ingredients.

In the spring I harvest cladodes from the *O. ficus-indica* growing in my yard (Fig. 1-1) that are about 1 month old and 15 centimeters (6 inches) in length or slightly longer. A potato peeler is used to slice off the raised areoles, those lateral buds with spines and the nasty glochids. I then slice the cladodes into thin strips about 6 mm (1/4 inch) wide (they can also be diced), which are simmered in water (sometimes with a little salt added) for a few minutes to blanche them and to let the unappetizing mucilaginous mucilage leak out. After decanting, I add fresh water, cilantro, oregano, garlic, and chopped onions and then cook them for a few minutes. The resulting product has a texture between green beans and okra. When the vegetable is served at dinner parties, everyone is amazed that I can cook, that they like nopalitos, and especially that they came from my yard.

To be efficient for the production of nopalitos, *O. ficus-indica* is often grown in clear plastic tunnels containing a few, closely spaces rows (Nobel, 1994; Sáenz-Hernández, Corrales-García, and Aquino-Pérez, 2002). The plants are heavily fertilized, usually with cattle manure or chicken manure. Annual production of nopalitos can be 400 tonnes per hectare (180 tons per acre), which is truly remarkable. After harvesting in the Mexico City area, the young cladodes are boxed, places in sacks, or more traditionally stacked in round columns 1.6 meters (5 feet) tall. The nopalitos are then taken to wholesale markets, where they are often distributed to street-side vendors. Using sharp knives and much skill, the vendors spend their time between sales scraping off the spines and glochids. Local people like to buy such fresh material. Increasingly, this operation is performed by machine, and the resulting nopalitos are enclosed in thin plastic wraps.

As already indicated, this vegetable is very versatile. Because nopalitos are high in fructose relative to glucose, such a vegetable is of great utility for people with type 2 diabetes mellitus (also called late- or adult-onset diabetes, it is the most common form of diabetes and is often simply called "diabetes"). Thus consuming nopalitos can avoid a major spike in blood glucose levels, which can occur when eating many other foods high in sugars, particularly glucose and sucrose, such as candies and cake. Although the daily intake of nopalitos recommended for such diabetics is high, such as 300 to 400 grams (11 to 14 ounces), the result is a much more even blood glucose level. Certain diabetics formerly needing insulin injections but switching to diets with high amounts of nopalitos can avoid such injections altogether. Consuming large amounts of many other leafy green vegetables can have similar effects, and more medical research is needed on this topic.

As indicated, the major country for nopalitos is Mexico, and the most commercialized production is near Mexico City. Altogether, about 8,000 hectares (20,000 acres) are used for nopalitos in Mexico. Besides the fresh market, a thriving business has developed for marinated canned (more properly "jarred") nopalitos. The jars usually contain vinegar, some spices, onions, carrot discs, and other vegetables. These can now be purchased in many countries and are ready to use in many dishes, such as scrambled eggs and omelettes. Yum!

Forage and Fodder—Cattle, Sheep, and Goats

To move from vegetables that we can eat—nopalitos—to what other animals can eat is a logical transition. In this regard, certain domestic animals used for meat, which we here restrict to cattle, sheep, and goats, outnumber humans more than ten-fold, and their diets can be controlled by humans. There is every reason to believe that they would all like platyopuntias, but often they have no choice in the matter.

At first I was startled by a feeding trial in Chile in the late 1980s. About half of the diet of penned cattle consisted of chopped cladodes of *Opuntia ficus-indica*—so far so good. The other half consisted of approximately equal amounts of chicken manure and alfalfa. Yuck! But the cattle seemed content, were well fed nutritionally, and gained weight. Another trial occurred in Texas, where cattle that also gained weight were fed cladodes of *Opuntia engelmannii*, bone meal (chopped, otherwise wasted bones from slaughter houses), and dried cotton seeds (cotton fiber was the first product, cotton seed oil expressed from the seeds was the second product, so this was the waste of the waste). These uses highlight the potential for platyopuntias as forage or fodder as well as the additional nutritional requirements needed for balanced diets.

Before further discussing such uses of *O. ficus-indica*, let us consider some other observations. When reviewing early uses, I mentioned that many wild animals feed on platyopuntias. In my backyard (Fig. 1-1), cladodes of *O. ficus-indica* are chewed by deer and ground squirrels. Yet adjacent plants of *Opuntia robusta* are untouched. Apparently taste matters.

But cactus spines are a problem. In Texas, the relatively flammable spines of the native *O. engelmanniii* are singed off using propane in backpack-mounted containers with blow-torch arms (Nobel, 1994; Felker et al., 2009). This releases food and water to Santa Gertrudis cattle (the first cattle breed developed in the United States). They love to eat the cladodes if they are free of spines. This is a great use of a local resource, as it can be withheld in wet times of plenty but made available when spines are burned off in dry times of need. A similar situation occurs for spineless *O. ficus-indica* developed by that great seed and plant merchant, Luther Burbank, at the beginning of the 20th century. He relentlessly touted his new and relatively expensive spineless cultivar in southern California. Now such spineless platyopuntias are raised all over the world as animal feed. When the virtues of platyopuntias are praised in later chapters, I am usually speaking of the spineless varieties, such as the one developed by Burbank.

Indeed, spines can be nasty, especially those with barbs along their shafts that tend to work their way into the skin. Such is the case for *Opuntia bigelovii*, whose common name is "jumping cholla" as the terminal segments sometimes mysteriously seem to jump onto a passing animal, including unwary humans. One noon in the Sonoran Desert of Arizona, I came upon a couple so afflicted. A segment of *O. bigelovii* had latched onto her hand. When she tried to remove it, her other hand became similarly impaled. He sportingly came to her aid, but soon both of his hands became impaled on the same segment. The more they struggled, the deeper into the skin the spines went.

They were going berserk as they stumbled hand-in-hand along the road, babbling incoherently about cacti with tears streaming down their faces. I stopped, calmed them down, gave them some water, cut off the spines with wire cutters, and drove them back to their car—much wiser to the ways of spines and cacti.

The area involved worldwide for cultivation of *O. ficus-indica* and other platyopuntias for forage and fodder now exceeds 1,900,000 hectares (4,700,000 acres; Nefzaoui and Ben Salem, 2002; Campos, Dubeux, and de Melo Silva, 2009; Nefzaoui and El Mourid, 2009), dwarfing all of the land uses for cactus fruits and nopalitos described above. The country with the greatest cultivation of platyopuntias for such purposes in Latin America is Brazil, with about 600,000 hectares (1,500,000 million acres) devoted to raising *O. ficus-indica* and *Opuntia cochenillifera* (*Nopalea cochinellifera*). The region is in the northeast, with a short rainy season followed by about eight months of drought and where freezing temperatures are uncommon. The harvested cladodes are fed mainly to cattle.

To find another country with an equally large area devoted to such production we must shift from the second largest country in the Western Hemisphere to a small county in northern Africa, Tunisia. Tunisia now also has about 600,000 hectares devoted to the cultivation of *O. ficus-indica* for forage or fodder, and the lucky recipient animals are mainly sheep. In Mexico, which has about 250,000 hectares (620,000 acres) of platyopuntias under cultivation for fodder (mainly for cattle), other animals that relish cladodes include not only goats but also horses and pigs. Indeed, anybody who carelessly leaves a gate open while traversing between plots of different uses in Mexico often inadvertently lets opuntias that are being raised for fruits or nopalitos be ravaged by hungry livestock.

The increased use of platyopuntias for cattle, sheep, and goats is perhaps their greatest promise for the future (Chapter 7). I remember talking to a cattle farmer in Senegal, who said that nothing eats the few cacti that he had planted. "Nothing?" He said, "O.K., goats love them!" Indeed, unconventional feed can be a secret for the increasing use of agaves and cacti in regions that receive enough rainfall during periods crucial for their survival and growth. But cladodes of platyopuntias are low in nitrogen, phosphorus, and sodium

(Nefzaoui and Ben Salem, 2002). Hence the unusual combinations of feed mentioned at the beginning of this subsection that provide nutritionally balanced diets. On the other hand, cladodes provide water, a crucial resource in arid and semi-arid regions. Importantly, cladodes can be the only source of water for livestock for many months at a time in various regions.

Cochineal—A Royal Dye

What could be more exciting than a cactus product that was once highly secret and worth more than gold! The color of this product varies with various factors, such as the acidity (pH) of a solution; it ranges from orange (at high acidity, low pH) to pink to red to a dramatic deep purple (most prized) as the pH is raised. The secret product is the cochineal dye produced by insects (Fig. 1-2). But let me tell you the story about cochineal in historical stages (see Nobel, 1994, 2002).

Figure 1-2. White webs of cochineal insects (*Dactylopius coccus*) infecting a cladode of *Opuntia ficus-indica* in Los Angeles, California. Pressing on a web to squeeze the underlying insect releases the beautiful red dye (carminic acid) onto your finger, which can be used to dye cloth, eggs, or other objects. [Plate I from Nobel, 1994; used by permission.]

Archeological evidence reveals the use of cochineal to dye woolen textiles that date to the Nazca and other cultures in Peru at the time of Christ. By the 10th century, the use of these dyes was common and widespread in Inca, Maya, and Aztec cultures. Indeed, Montezuma, the great Aztec emperor, had a magnificent, deep purple robe. His robe led to awe among his subjects, who were forced to supply a tax to the emperor in the form of dead insects. Not any insects, but special ones that contained a vivid red dye—the cochineal insects that fed on the platyopuntias *Opuntia ficus-indica, Opuntia tomentosa, Nopalea cochenillifera*, and a few other related species.

As larvae, female cochineal insects of *Dactylopius coccus* (and a few other species) crawl onto a cladode, insert their mouth parts into the stem, and spin a characteristic white web around them for camouflage (Fig. 1-2). They then suck nutrients from the plants for up to three years. Such sessile parasites have preyed upon many opuntias in my yard, and they are very hard to eradicate, but back to the story. First, for sexual equality, I note that the males are smaller, winged (to get to where the action is), and live only a few months.

In the 16th century, Spanish conquistadors in present-day Mexico were entranced by the colorful robes of the Aztecs. They began sending the dried royal dye back to Spain. Later they arranged to have the platyopuntias on which the female insects fed sent to Europe. An industry developed in the Canary Islands, which was shrouded in mystery so that nobody could compete with the entrepreneurs. The brilliant Dutch scientist who refined microscopes and gave us the first clear definition of cells, Anton van Leeuwenhoek, solved this mystery in 1704. He noticed the unmistakable insect parts in the cochineal dye that he employed, which at the time was used as a stain on microscope slides. However, the dye was much more famous for staining clothes.

The vibrant but expensive cochineal red dye was prized by European royalty. Look at the wonderful portraits in museums worldwide of Spanish, Portuguese, French, and British kings, queens, counts, and other dignitaries of that era. When Paul Revere warned in 1775 that the "redcoats were coming," he was referring to the striking red jackets of the British Regulars, dyed with the special ingredient from cochineal insects. The tradition was continued by the Canadian Mounted Police ("Mounties," now known as the Royal Canadian Mounted Police).

In the 18th century, cochineal dye was second in value only to silver as an export from Mexico. However, in the latter half of the 19th century synthetic dyes produced from coal tar, which were stronger, faster binding, and much less expensive, basically killed the cochineal industry. A resurgence occurred in the early 20th century, as coal-tar dyes were linked to certain medical problems including cancer, so carminic acid (the main coloring agent in cochineal) and various other natural dyes were certified for use as food colorants. Now the dye finds uses in dying crab, in curries, in coloring the Italian aperitif Campari (ingredients a secret since 1860), in soft drinks, in cosmetics such as lipstick, in color printing, in bacteriological and other biological stains, as well as a color prized by artists (carmine, also known as Crimson Lake).

Let us next consider some of the details of the modern production of carminic acid. It has the incredible, basically unpronounceable, and hard to remember chemical name of 7-α-D-glucopyranosyl-9,10-dihydro-3,5,6,8-tetrahydroxy-I-methyl-9,10-dioxo-2   anthracene carboxylic acid! Carminic acid, which has a molecular weight of 492, is also known as Natural Red 4, Color Index 75470, and E120. To produce the dye, young peripheral cladodes (often about one year old) of *O. ficus-indica* can be detached, placed in a shed, and then infected with *Dactylopius coccus*, which occurs in the superfamily Coccoidae that includes mealy bugs and scale insects. When mature, the female insects are 3 mm (⅛ inch) or slightly longer. If you squash the cottony white web on a cladode, a beautiful purplish stain occurs on your fingers (Fig. 1-2), which you can use to adorn your face, your clothes, or anything nearby.

Not much dye occurs in a single insect. That is why approximately 130,000 adult female insects must be collected to produce 1 kilogram (2.2 pounds) of the dye. On a good day, the dye constitutes about 20% of the dry weight of the insects' bodies. Based on ancient tradition, it is usually extracted using hot water. Although mechanical harvesting is today the norm (usually by blowing air across the cladodes), instead of hand picking the insects off of the cladodes that is also practiced, the dye is still expensive—not approaching the price of gold, as was the case in the 16th century, but still expensive.

Currently, the main producing country for commercial carminic acid is Peru, with some production in Mexico (especially in the state of Oaxaca), the Canary Islands, Argentina, Bolivia, Chile, Ethiopia, Morocco, South Africa, and a few other countries. Today aproximately 600 tons are produced annually. Growing *O. ficus-indica* to support the cochineal industry requires about 25,000 hectares (60,000 acres) worldwide. As indicated, *D. coccus* has preyed upon many opuntias in my yard (Fig. 1-2). Indeed, it and closely related species have been used as a biological control agent to limit the invasiveness of certain platyopuntias, especially in Australia and South Africa. This invasiveness by platyopuntias is not really tragic, but rather a testimony to the great potential for biomass productivity by *O. ficus-indica*, as will be discussed in due course.

Peyote and other Hallucinogens

Although they do not have much to do with global climate change, various psychoactive compounds in cacti are interesting. The most famous is peyote, which is obtained from *Lophophora williamsii* (commonly called *peyote*). One of my friends had a cat called "Lophophora." Only after I learned some cactus taxonomy did I get his drift … hmmm. Also, when someone donated a cactus collection with 200 species to a university in the western United States, it accepted every one except *L. williamsii*, as the university did not have enough security to prevent its inevitable theft.

This small, essentially spineless cactus is native to northern Mexico and southern Texas. When mature, it is about 80 millimeters (3 inches) across, with little of its rounded stem above ground level. After collecting and drying, the ingested stem or flowers lead to mental exhilaration caused by mescaline. The resulting colorful hallucinations are ritualistically (aka, religiously) experienced by about 200,000 believers from Mexico to Canada. It is legal for Native Americans to use in the United States, but a federal permit is required for its possession. Such use of peyote has occurred for over 2000 years (Anderson, 1980).

Mescaline is chemically related to other compounds that affect the transmission of nerve impulses, such as serotonin. Likewise, it stimulates the sympathetic nervous system, increasing blood pressure, dilating pupils, and leading to arousal patterns in the brain. Besides *L. williamsii*, mescaline is also present in the stems of *Opuntia cylindrica* from Ecuador and Peru. In addition, these latter two countries also have San Pedro cactus (*Trichocereus pachanoi*), which reputedly has more mescaline than *L. williamsii* per unit biomass. Various other cacti contain many related compounds that also have hallucinogenic properties.

From Medicine to Sweetness

Peyote can be considered to be a "medicine." Besides its spiritual effects, it has been ground up and applied to wounds as well as ingested to cure sickness and to reduce internal muscle spasms (Anderson, 1980). Just as for many industrial chemicals occurring in agaves discussed above, including compounds that can act as hormones for livestock, cacti contain an untapped wealth of chemicals waiting to be discovered. Indeed, various steroids can be extracted from many columnar cacti. Although we must not lose our focus on the big picture with respect to productivity and global climate change, many similar smaller uses of cacti are currently emerging.

For example, the stems of many species of barrel cacti can be infused with sugar solutions to produce "cactus candy." But this leads to the unnecessary harvesting of such beautiful cacti, which are relatively slow growing. On the other hand, cactus pears and even nopalitos are used in jams and marmalades. Cactus pears are also used to color and sweeten soft drinks and are fermented to produce alcoholic beverages, mainly in Mexico. In addition, because of their sweetness and attractive colors, cactus pears are used in ice creams and sorbets. Although these facets have not yet reached their commercial potential, it is a logical and admirable extension of traditional uses. Mucilage, which is removed to produce edible nopalitos (see above), also has value. This complex polymer of sugars can be utilized as a food thickener and as an adhesive. Let us salute the cactus entrepreneurs, as their vision will create new jobs and greater opportunities for these plants in the future.

And again Ornamentals

As for agaves, the love of cacti as house and yard plants has inspired enthusiasts and collectors all over the world. How many times have I visited a home in northern or eastern Europe, or in Japan, to be shown some of the most incredible specimens of *Mammillaria* that I have ever seen. Their many diminutive and beautiful red flowers on plants in south-facing window boxes add cheer in the spring and hope for the upcoming summer. Other popular small cacti include many species of *Rebutia*, with their relatively large and dazzling red, yellow, or pink flowers. Moreover, if one forgets to water these plants for weeks that extend into months, they do not complain.

Larger cacti are also used as ornamentals. Witness the striking barrel cactus with golden spines, *Echinocactus grusonii*, which is humorously known as "mother-in-law's seat." Another favorite is *Cephalocereus senilis* ("old man cactus") with its long, flexible white spines. Branching columnar cacti are also popular in gardens, such as *Cereus peruvianus* (*Cereus repandus*) and *Myrtillocactus geometrizans*, both of which have edible fruit.

Then there is the stately and symbolic *Carnegiea gigantea* (commonly known as *saguaro*). It is the state flower of Arizona and is featured in many western movies and photographs. Besides their dramatic shape and beautiful flowers, cacti make ideal ornamentals because they require little attention with regard to fertilizer application and especially watering. Hence they are treasured in yards and botanical gardens around the world. Cacti are even recruited as attractive security agents, being planted as spiny fences to protect properties and under windows to deter intruders.

What to Expect in Future Climates—A Preview

This first chapter has covered the main current uses of agaves and cacti. In the final chapter, which for many will be the most important in the book, we will consider future shifts in such uses in response to global climate change. But first we develop some underlying principles.

We begin with the special photosynthetic pathway used by agaves and cacti that leads to water conservation (Chapter 2). Next, we present their excellent tolerances of drought and of high temperatures but notoriously poor tolerance of low temperatures (Chapter 3). After quantifying the main climatic changes expected in the 21st century—rising atmospheric CO_2 levels, rising temperatures, and changing rainfall patterns—we will relate such changes to the survival of agaves and cacti (Chapter 4). We then explore how their CO_2 uptake is influenced by temperature, light, soil water and nutrient status, and the atmospheric CO_2 level (Chapter 5). From net CO_2 uptake to the productivity of agaves and cacti is a short step, but it must be done carefully (Chapter 6). We are then ready to evaluate these remarkable plants with respect to the future (Chapter 7), a primary objective of this book. Our considerations in this final chapter will embrace fodder, biofuels, erosion control/desertification, and carbon sequestration, including the use of carbon credits.

As we shall see, the net CO_2 uptake ability and the biomass productivity of agaves and cacti will be enhanced by the increasing atmospheric CO_2 concentrations associated with global climate change. The increasing temperatures will expand the possible areas for their cultivation. Currently, fodder from opuntias represents the largest land use of agaves and cacti worldwide, and the use of fodder has incredible potential for major expansion in the future. Please be patient as the background is developed before the major and positive take-home lessons are presented. But if you are impatient, go ahead and jump to Chapter 4 to see the details and consequences of climate change. And then jump to Chapter 7, where future applications and regional implications are considered.

2

Advantages of Crassulacean Acid Metabolism

Three Photosynthetic Pathways—Ecology

Everyone has a general feeling about what photosynthesis is. Based on "photo-", it obviously involves light. The "synthesis" part involves the production of something, such as a carbohydrate. We can represent the general chemical reaction for photosynthesis in its simplest but correct form as

carbon dioxide plus *water* reacts to form
    carbohydrate plus *oxygen*   (2.1a)

or, in chemical symbols

$$CO_2 + H_2O \leftrightarrow CH_2O + O_2 \qquad (2.1b)$$

Photosynthesis is the largest synthetic process on earth. According to some estimates, the annual photosynthetic productivity is about 110 $\times 10^{12}$ kilograms (110,000,000,000 tonnes) of carbon converted into photosynthetic products each year (we note that 10^{12} kilograms = 10^{15} grams = 1 petagram, a unit often used in world carbon budgets; Chapin et al., 2002; Nobel, 2009).

The light involved in photosynthesis generally comes from the sun, but also it can be supplied by tungsten or fluorescent lamps. The advantage of fluorescent lamps is that about 90% of their energy is in the visible region, where pigments such as chlorophyll absorb. On the other hand, tungsten lamps emit about 90% of their energy in the longer-wavelength infrared, which photosynthetically speaking is not useful. This was not appreciated by some growers of a thin-leaved plant that is used medicinally in a few states and illicitly used everywhere recreationally (you know the plant I mean). To keep their operation covert, they grew the plants in their basement, but stupidly used extravagant banks of tungsten lamps. The local power company saw their huge consumption of electricity, came to investigate, and noticed a telltale odor wafting from the house. Had the growers used fluorescent lamps with nearly ten times as much visible light per dollar, they would have been happily undetected. But let us get back to photosynthetic pathways.

As presented in this and the next section, three different photosynthetic pathways occur in plants, and soon you will be familiar with all three of them. The impact of biochemistry in elucidating these three photosynthetic pathways owes a lot to the technology developed during World War II (Black and Osmond, 2003; Nobel, 2009). In the war effort, various radioisotopes were produced (such as ^{14}C), which later were crucial for tracing metabolic pathways. Also, instruments such as spectrophotometers and magnetic resonance imaging (MRI) devices were perfected (currently used extensively in hospitals). And techniques such as electrophoresis and chromatography were improved. But before this era, the three photosynthetic pathways had actually been stumbled upon much more humbly.

In 1682, Neimiah Grew reported the special acidic (bitter) taste of succulent plants in the morning. In 1804, Nicolas–Théodore de Saussure investigated the nocturnal uptake of CO_2 by such plants. Also in the early 19th century, a gentleman in India systematically bit into his cacti and other succulent plants at dusk and then again at dawn. As reported by Benjamin Heyne in 1815, they tasted much more acidic at dawn. This was an early if mystifying demonstration of the now explainable buildup of acid in Crassulacean Acid Metabolism (CAM) plants at night.

In the early part of the 20th century, the amount of biomass accumulated per amount of water lost was found to vary among plants. Careful scrutiny of the data shows that some plants accumulated about twice as much biomass per water lost as other plants under the same environmental conditions. The ones that were more efficient in using water are now known as C_4 plants and the others are known as C_3 plants ("C" stands for carbon; the subscript will be explained shortly). Before discussing the biochemical differences among C_3, C_4, and CAM plants (this trio of photosynthetic pathways encompasses all plants), let me summarize some of the ecological differences.

Climates vary—and for plants to be successful, they must adapt to their environments. Today slightly over 90% of the approximately 300,000 currently recognized species of vascular plants use the C_3 photosynthetic pathway. Thus, with an impressive air of certainty, you can point to any plant and say: "Hey, that's a C_3 plant!"

To identify plants using the other two photosynthetic pathways, you should seek out those in extreme environments and special *microhabitats* (small, local sites) or plants that are growing rapidly. A particular case is weeds. Indeed, 8 out of the 10 agriculturally most noxious weeds use the C_4 pathway. You should also look for plants producing extremely well, such as corn (maize, *Zea mays*), sorghum (*Sorghum bicolor*), and especially sugarcane (*Saccharum officinarum*). These are also C_4 species. Another characteristic of C_4 species is substantial growth under high light and high temperature. Thus, many tropical and subtropical grasses are C_4 plants. Despite their high proportion among weeds and highly producing species, only 1 to 2% of vascular plant species use the C_4 pathway.

Now we come to CAM plants, such as agaves and cacti, the focus of this book. Deserts can offer extreme conditions in terms of water stress and high temperature. In the New World, this is where CAM plants often reign supreme, because they are the champions at conserving water. Look at the magnificent columnar and barrel cacti of the southwestern United States and much of Latin America. In the Old World, look at the tall succulent euphorbias of South Africa, Zimbabwe (if you can get there), Botswana, Mozambique, and Madagascar. Some of these plants are growing in arid regions, but most are growing in semi-arid regions (such regions are defined in Chapter 1; also see the Glossary).

As striking as most cacti and agaves are, with their impressive, dominating features, most CAM plants are far less conspicuous. Actually, most CAM species are native to relatively wet tropical and subtropical regions but occur in very dry microhabitats. For example, there are many CAM epiphytes growing along branches and in crotches of trees, where water is relatively scarce even a few days after a rainfall. Witness the beautiful orchids with their enchanting flowers of all kinds of shapes and colors in tropical rainforests. Witness the bromeliads of tropical and subtropical occurrence, often with showy flowers. Bromeliads range from the diminuitive *Tillandsia usneoides* (Spanish moss) draped from telephone wires and trees to species of *Aechmea, Brocchinia,* and especially *Vriesea,* whose large leaves [often over 0.8 m (2½ feet) long] dominate the tops of certain trees in Amazon rainforests. All told, 6 to 7% of vascular plant species use the CAM pathway, most of which are epiphytes or hemiepiphytes.

Although agaves are generally not such opportunistic plants, various opuntias have been seen growing on castle walls in Europe and on archeological ruins in Latin America. Moreover, 130 species of cacti are true epiphytes or hemiepiphytes that are found in neotropical rainforests and woodlands, such as species of *Rhipsalis.* This genus includes *Rhipsalis baccifera,* the most widespread of all cactus species and the only one native to the Old World (perhaps because its sticky fruit was widely dispersed by birds). It is readily found in forests of tropical Africa, Madagascar, and Sri Lanka. For CAM plants, the 'raison d'etre' is biomass production for a limited amount of water used, as we will consider in this chapter and the last two chapters.

Two Key Enzymes—What, Where, and When

Now to the nitty gritty of the three photosynthetic pathways. The keys are two enzymes, which have long names and pivotal roles. One is called ribulose-1,5-bisphosphate carboxylase/oxygenase, which we will shorten to "Rubisco." The other is called phosphoenol pyruvate (PEP) carboxylase, which we will shorten to "PEPCase." Rubisco is the most prevalent enzyme on Earth — about 10 kilograms (22 pounds) of Rubisco occurs in plants for every person! PEPCase is also common, perhaps a kilogram or two per person.

What these enzymes do is remarkable, but also rather disappointing in the case of Rubisco. First, what is an enzyme? An enzyme is a protein that facilitates a specific biochemical reaction, where "facilitates" means speeding it up, usually by a factor of one thousand to many millions. Thus changes that are rather improbable, such as fixation of carbon dioxide into a carbohydrate (Equation 2.1), become much more probable in the presence of the right enzymes. Selecting or favoring certain reactions or processes among the many that are possible is the biochemical essence of living organisms.

Rubisco takes a five-carbon compound (ribulose-1,5-bisphosphate) and binds CO_2 coming from the atmosphere to it. The resulting six-carbon compound then splits into two *three*-carbon compounds (3-phosphoglycerate), which is the basis for the number three in the name C_3 photosynthesis. PEPCase also binds CO_2 but this time onto a three-carbon compound, phosphoenol pyruvate (PEP); this yields a *four*-carbon compound, such as malate or oxaloacetate (all of the compounds mentioned are carbohydrates). Hence the number four occurs in the name for this pathway, C_4 photosynthesis. Science is complicated, as three biochemically distinct C_4 pathways exist, but the end result is the same for each of them in terms of CO_2 being stored as a carbohydrate (Equation 2-1). This is the "What" for C_3 and C_4 plants (CAM plants will be considered shortly).

The "Where" is also interesting for C_3 and C_4 plants (Fig. 2-1). Rubisco occurs in chloroplasts, the location for chlorophyll and hence for photosynthesis. These chloroplasts occur in *mesophyll* cells for C_3 plants, where "meso" refers to middle and "phyll" refers to leaf; thus mesophyll cells are the green cells that occur within a leaf. Turning now to the mesophyll cells of C_4 plants, we note that PEPCase occurs in their *cytosol*, which is the region of a cell outside of organelles such as chloroplasts and where other biochemical reactions besides photosynthesis take place. The fact that PEPCase binds the hydrated form of CO_2, bicarbonate, is interesting but rather incidental to the consequences for carbon sequestration by plants.

More important for "Where" is the anatomy. And as we shall see, compared with C_3 plants, C_4 plants have an anatomical step added before CO_2 coming from the atmosphere is fixed into photosynthetic products.

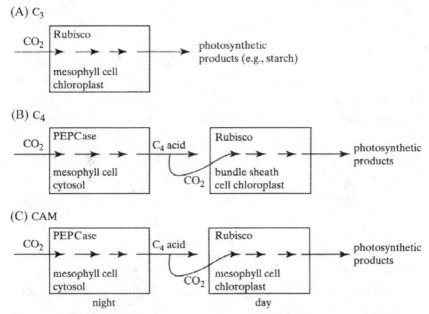

Figure 2-1. Roles played by the two key enzymes, Rubisco and PEPCase, in the three photosynthetic pathways: (A) C_3, (B) C_4, and (C) CAM. The reactions for the C_3 and the C_4 plants occur during the daytime, when their stomata are open. Stomatal opening for CAM plants occurs predominantly at night; during the daytime the stomata of CAM plants close, so the CO_2 released inside the leaves or the stems does not easily leak out then. [Figure 8-15 from Nobel, 2009; used by permission.]

C_4 plants were early recognized but not biochemically distinguished from C_3 plants by their more efficient use of water, as mentioned above. In retrospect, they were also recognized by a peculiar anatomy, where the small veins in leaves appeared very dark green. This "Kranz anatomy" is due to bundle-sheath cells (mentioned in Fig. 2-1) surrounding the vascular tissue that contain many chloroplasts. Indeed you could say: "Hey, this is a C_4 plant!" if you hold up to the light leaves of corn (maize), sorghum, or sugarcane and notice the dark-green lines of the veins. The analogous cells in C_3 plants contain few chloroplasts and are thus not visually prominent. But the many chloroplasts in the greenish bundle-sheath cells of C_4 plants contain much Rubisco and are the sites where the photosynthetic products are made (Fig. 2-1).

Now let us attack the "When" of the photosynthetic enzymes (Fig. 2-1). C_3 plants fix CO_2 into carbohydrates via Rubisco in the chloroplasts during the daytime, when the stomata are open (Fig. 2-1). ('Stomata' is often anglicized to 'stomates,' although most scientists prefer 'stomata.') Stomata are the tiny pores in leaves that open to allow CO_2 to enter, which is necessary for *photosynthesis*. Stomatal opening, however, leads to the loss of water vapor through the same pores, which is known as *transpiration*. C_4 plants make it anatomically and biochemically more complicated than C_3 plants. They fix CO_2 via PEPCase in the cytosol of mesophyll cells during the daytime, then shuffle the four-carbon acids to the bundle-sheath cells, then decarboxylate the four-carbon acids thereby releasing CO_2, which then diffuses into the chloroplasts of bundle-sheath cells, where it is fixed into photosynthetic products by Rubisco, all during the daytime (Fig. 2-1). Whew! This scheme encompasses the extra step referred to above for C_4 plants compared with C_3 plants.

CAM plants are rather extraordinary, and both key photosynthetic enzymes figure prominently in their biochemistry (Fig. 2-1). Biochemically, CAM plants closely resemble C_4 plants, but they differ in the "Where" and especially the "When." They use PEPCase to fix most of the CO_2 into four-carbon compounds in the cytosol, but at *night*. This requires that the stomata of CAM plants must be open at night to let the CO_2 into the leaves or stems. But there is no light at night and hence no photosynthesis then. To capture or "fix" the CO_2, the CO_2 molecules entering CAM plants are bound in the form of organic acids (mainly malic acid, but also oxaloacetic acid and citric acid). This causes the pH to decrease as the acidity increases, as noted in the early 19th century by the gentleman biting his cacti.

The uniqueness of the "When" for CAM plants continues. The bound CO_2 is released the next day within the plants as the sun comes up (or the lights go on in a house or in an environmental chamber for growing plants). This CO_2 is then available to Rubisco (Fig. 2-1), which now performs photosynthesis in the light, as usual. The stomata of CAM plants are mostly closed during the daytime, which prevents the escape of the internally released CO_2. Thus the spatial (cellular) separation between cells for the initial CO_2 fixation and for the Rubisco in C_4 plants is replaced by the temporal separation within the same cell in CAM plants (Fig. 2-1). The timing of this biochemistry has major consequences for the efficient use of water considered later in this chapter.

So far we have not explained the term CAM, other than to say it is an abbreviation for *Crassulacean Acid Metabolism*. The name comes from the fact that this pathway was observed and carefully studied in the family Crassulaceae, which contains the leaf succulent jade plants and sedums. What is the most famous CAM plant? The prize goes to *Ananas comosus*, pineapple! This bromeliad is enjoyed as a fruit worldwide and is cultivated on approximately 900,000 hectares (about 2,200,000 acres) in various countries. Initial hints about the uniqueness of CAM were based upon the nocturnal stomatal opening, which, as we will see, is the most curious aspect of CAM and also the most important aspect for water conservation. Also, the tissue acidity of CAM plants increases at night, which accounts for "Acid Metabolism." Now we know the reason for the acronym "CAM."

Before concluding this discussion of the enzymes involved, let us note a few dates and facts (Winter and Smith, 1996; Black and Osmond, 2003; Osmond et al., 2008; Nobel, 2009). Tracing the biochemical events of C_3 photosynthesis involved the availability of ^{14}C (a radioactive isotope of carbon) and improved chromatographic techniques, as evidenced by the pioneering work of Melvin Calvin (Nobel Prize in Chemistry in 1961), Andrew Benson, and James Bassham in California in the 1940s. Much of the biochemical essence of the CAM pathway was described at about the same time and on into the 1950s by Harry Beevers, Stanley Ransom, and Meirion Thomas working in England and Hubert Vickery working in Connecticut, among others. However, the relationship of CO_2 uptake at night to photosynthesis for CAM plants was not clearly understood at that time. In the 1960s, Hal Hatch and Charles Slack in Australia provided evidence for the C_4 photosynthetic pathway.

Biochemical control mechanisms and molecular details of all three photosynthetic pathways are still being elucidated.

Yet a character flaw of the most common enzyme, Rubisco, is its propensity to bind O_2 at the same site that it binds CO_2. Binding O_2 causes the eventual release of CO_2, which can return to the atmosphere. This is termed *photorespiration*, because it occurs in the light (hence "photo-") and releases CO_2, as does respiration. Photorespiration undoes the accomplishment of photosynthesis. This tarnishes the magnificence of the otherwise extraordinary protein, Rubisco, and shows that evolution is not always perfect. Also, photorespiration is an important feature of the biochemical consequences of global climate change with respect to increasing CO_2 concentrations and increasing temperatures.

The current pace of environmental change is so rapid that most plant species do not have enough time to adapt, except possibly for those with very short generation times. In this regard, photorespiration (the bad guy) increases faster with increasing temperature than does photosynthesis (the good guy), which can have serious consequences as the global temperature rises. For instance, temperatures could increase so much in some regions that less net CO_2 uptake might occur, where net CO_2 uptake is photosynthesis minus respiration plus photorespiration, a topic that we will consider in Chapter 5 for agaves and cacti.

Nuts and Bolts of Gas Exchange

To assess plant behavior with respect to climate, we must measure CO_2 uptake by photosynthesis, a process represented in Equation 2.1. That equation also indicates that O_2 is produced by photosynthesis. At the current rates, such oxygen evolution by plants and other photosynthetic organisms (algae and some bacteria) replenishes the entire atmosphere's oxygen content in about 2,000 years.

With respect to oxygen, everyone knows that bringing flowers to people in hospitals will cheer them up. Some believe that the flowers and the accompanying green leaves will supply the patients with O_2. This is actually not true, as hospital rooms are too dark to lead to much photosynthesis during the daytime (except near a window or a bright light). At night, the flowers and the leaves of the bouquet are actually taking up O_2 needed for respiration, which occurs for all living organisms all of the time. Over a 24-hour period, the colorful and thoughtful gift actually leads to less oxygen in the hospital room. But do not worry, as the effect is vanishingly small. Specifically, the oxygen concentration in the atmosphere is approximately 210,000 parts per million (21%), while the CO_2 concentration is only about 400 parts per million (discussed in the first section of Chapter 4).

The O_2 concentration is rarely measured by plant biologists in gas exchange studies (it does not change appreciably). But the concentrations of water vapor and CO_2 nearly always are. Loss of water vapor by plants—transpiration—is an inevitable consequence of the stomatal opening necessary for CO_2 uptake to support photosynthesis. Thus to discuss plant gas exchange, we will focus on CO_2 and H_2O.

Figure 2-2 shows the "nuts and bolts" of gas exchange measurement for plants. The idea is to isolate a leaf from the surrounding air and to place it in a chamber. We then measure the levels of CO_2 and water vapor in the air entering the chamber and compare these measurements with the levels for CO_2 and water vapor in the air exiting the chamber. Any differences in the two measurements are due to gas exchange by the leaf. If the exiting CO_2 concentration is lower than the entering one, then net CO_2 uptake has occurred by photosynthesis. If the exiting CO_2 concentration is higher, then CO_2 release by respiration (and also by photorespiration, described above) is greater than the CO_2 uptake by photosynthesis. CO_2 release occurs under low levels of illumination for C_3 and C_4 plants during the daytime and for all C_3 and C_4 plants at night. CAM plants are more complicated, as we will see shortly.

Although the CO_2 level in the exiting air can be higher or lower than that in the incoming air, the water vapor concentration is always higher in the exiting air. Why? If the stomata are open, then the high water vapor concentration in the leaves leads to water vapor diffusing out of them to the surrounding air, which in general has a much lower water vapor concentration. Transpiration! Even when the stomata are closed, however, the high water vapor concentration in plants leads to water loss.

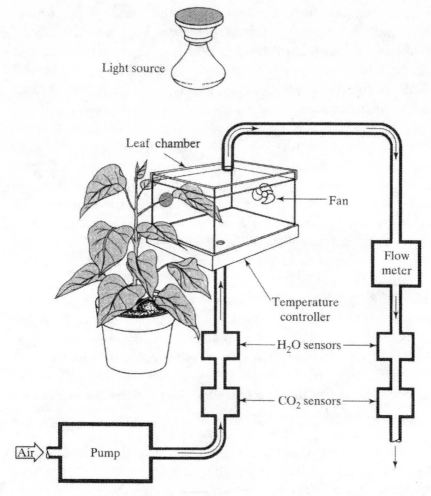

Figure 2-2. Schematic version of the equipment used to measure gas exchange for an attached leaf. Modern instruments use the same principles but are less bulky, battery powered, and hence portable; the light source can be the sun. Correctly mimicking local wind speeds, here done by a fan, can be a problem.

As an indication of a transpiration calculation, let us suppose that the air entering the chamber contains 0.6 mole of water vapor per cubic meter and that the exiting air contains 1.0 mole of water vapor per cubic meter—the increase in water vapor concentration is due to the leaf's transpiration. We will consider a relatively slow air flow rate of 1.0×10^{-4} cubic meters per second (100 cubic centimeters per second). Let us assume that the leaf has a representative surface area of 1.0×10^{-2} square meters (100 square centimeters, or 15 square inches). The transpiration rate then is $(1.0 \text{ mol m}^{-3} - 0.6 \text{ mol m}^{-3})(1.0 \times 10^{-4} \text{ m}^3 \text{ s}^{-1})/(1.0 \times 10^{-2} \text{ m}^2)$ or $0.004 \text{ mol m}^{-2} \text{ s}^{-1}$, which is a typical transpiration rate for the leaf of a C_3 plant. [Figure 8-1 from Nobel, 2009; used by permission.]

Such water loss with closed stomata is referred to as *cuticular transpiration*, as it occurs across the water-proofing waxy layer on leaves that is referred to as a *cuticle*. This cuticular transpiration cannot be stopped. But as we shall see, agaves and cacti have adaptations that can greatly reduce this water-losing "downer." Such adaptations include (1) a much thicker cuticle than for plants growing in regions or microhabitats with more soil water availability, and (2) stomata that close extremely tightly, thus shutting off essentially all of the gas exchange through them.

Figure 2-2 shows a leaf in a chamber. Other plant parts, and indeed a whole plant, can also be put in such a chamber. At an even larger scale, the chamber concept can be adapted to groups of plants—again, both the incoming and the outgoing water vapor and CO_2 concentrations are monitored. Figure 2-2 is reminiscent of the apparatus that various plant ecophysiologists built in the uniquely creative 1960s and 1970s. During that period plant physiological ecology, meaning making and recording physiological measurements for plants in the field and drawing ecological conclusions, came of age under pioneers such as Dwight Billings and Hal Mooney. Measurement of plant gas exchange with better and better instruments was a key to this emerging discipline, which is crucial for understanding plant photosynthetic performance in the field.

The commercial potential of gas exchange measurement was soon realized by entrepreneurial companies, and such instruments became available for purchase. Moreover, use in remote locations demanded portability. The sophisticated current models can cost $40,000 (U.S. dollars), or much more than the cost of an economy car but with no horn, no tires, no radio, and no creature comforts! Although these new instruments have control of temperature and light, advanced software, and other features, certain aspects are age-tested. For example, CO_2 is usually measured based on its strong absorption of longwave (thermal) or infrared radiation. Indeed, this is related to the warming associated with the globally increasing CO_2 levels in the atmosphere (discussed in Chapter 4). Various infrared sensors and solid-state devices can accurately measure water vapor and determine the relative humidity, which has been a major improvement in the instruments over the years.

Relative humidity is an important parameter in plant studies. It represents the water vapor content of the air compared to the maximal or saturation water vapor content of the air:

$$\text{relative humidity (in \%)} = \frac{\text{air water vapor content}}{\text{saturation (maximal) water vapor content}} \times 100$$

(2.2)

Relative humidity is very temperature dependent, as we shall see soon. It is usually expressed as a percentage, as in Equation 2.2. For example, 50% relative humidity means that the air contains half of the maximal or saturation concentration of water vapor at that temperature.

Besides being able to measure and to vary relative humidity, some modern gas exchange instruments can also measure and vary the CO_2 level in the leaf chamber, which is very pertinent to global climate change. We can then see how a leaf responds to atmospheric CO_2 levels over the short term. Understanding longer-term responses to atmospheric CO_2 levels is more complicated and will be considered in Chapter 5 for agaves and cacti. In any case, the great increase in the sophistication of the gas-exchange instruments has greatly increased our ability to make the requisite measurements of net CO_2 uptake and water vapor loss for plants in the field.

Daily Gas Exchange Patterns

Now that we know in principle how to measure net CO_2 uptake, let us consider some data for agaves, cacti, and other plants. In the good old days, such measurements were done with home-made instruments and were not for everyone. Our biological rhythm coaches us to sleep at night and to be active during the daytime—but anyone who knows cats also knows that their daily rhythm is different from that of most of ours. Specifically, CAM plants, like cats, tend to be active at night—some say that they are "working the night shift" (Black and Osmond, 2003). By *active* at night is meant that such plants take up CO_2 then. Cats release CO_2 at night, as do over 90% of plant species, namely, the C_3 and the C_4 types described above.

To understand fully the daily life of plants, we need to make gas exchange measurements over 24-hour periods (Fig. 2-3). This is routinely done for CAM plants. But such measurements are instructive for plants of all of the three photosynthetic pathways. Let us begin with the predominant plant species on Earth, those using the C_3 pathway. Their stomata behave as expected and open during the daytime. CO_2 uptake therefore rises rapidly at dawn, as Rubisco in the mesophyll cells then leads to the enzymatic binding of CO_2 that leads to carbohydrate formation (namely, photosynthesis; Equation 2.1). Sometimes with a reversible decrease near noon, CO_2 uptake by C_3 plants abruptly decreases at dusk (Fig. 2-3), when the low light level greatly reduces photosynthesis. At night, there is no light, hence no photosynthesis, and no CO_2 uptake but rather a CO_2 release by C_3 plants. C_4 plants have a similar daily pattern. Namely, their CO_2 uptake occurs during the daytime, with a rate that on average is somewhat higher than for C_3 plants, and CO_2 release again occurs at night (Fig. 2-3).

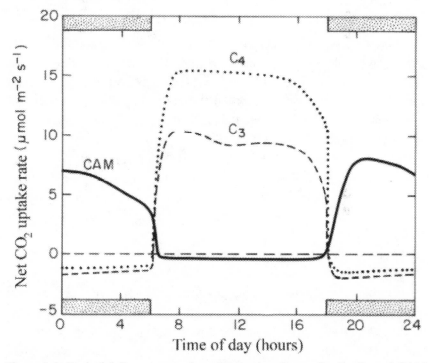

Figure 2-3. Typical daily patterns of net CO_2 exchange rates by C_3, C_4, and CAM plants under good growing conditions. Higher maximal rates would occur for rapidly growing C_3 plants (e.g., lettuce, spinach, and wheat), rapidly growing C_4 plants [e.g., corn (maize), sorghum, and sugar cane], and rapidly growing CAM plants (e.g., *Agave mapisaga*, *Agave tequilana*, and *Opuntia ficus-indica*). Lower maximal rates of net CO_2 exchange would occur for most C_3 and C_4 trees and for

slow growing CAM plants, such as *Mammillaria* spp., which are often only 0.20 meters (8 inches) tall at 50 years of age. The stippled region indicates the night. [Figure 2-1 from Nobel, 1988; used by permission.]

Speaking of rates, let me comment on the units for photosynthesis. Scientists use the metric system, such as that in Figure 2-3, where the units for net CO_2 uptake by plants are generally $\mu mol\ m^{-2}\ s^{-1}$. This stands for micromoles (μmol, where μ is the symbol for 10^{-6} and mol is the shorthand for mole, which is an amount of something containing Avogadro's number or 6.02×10^{23} molecules or other particles) of CO_2 taken up per square meter (m^{-2}) per second (s^{-1}). This does not mean that leaves are a few meters in area or that the measurements are made only over a few seconds. This reminds me of a person pulled over by a policeman who said "You were speeding at 60 miles per hour!" To which the person replied, "No, I have only been driving for 2 minutes!"

Now to CAM plants. In contrast to C_3 and C_4 plants, their net CO_2 uptake occurs predominantly at night (Fig. 2-3). The rate of CO_2 uptake usually increases rapidly just after dusk, when their stomata open, reaches a maximum early in the night (at least under wet soil conditions), and then decreases throughout the rest of the night. During most of the daytime, the stomata of CAM plants close tightly, so little net CO_2 exchange with the environment occurs then. The slight net release of CO_2 by CAM plants during the daytime implies that the CO_2 concentration within the leaves of agaves or the stems of cacti is higher than the CO_2 concentration in the air surrounding these plant parts. This higher internal CO_2 concentration in CAM plants during the daytime is a consequence of the internal release of CO_2 from the acids that accumulated during the previous night (Fig. 2-1).

Ecologically speaking, the key for the consequences between predominantly nocturnal gas exchange by CAM plants and predominantly daytime gas exchange by C_3 and C_4 plants is temperature. Namely, temperatures are lower at night, and lower temperatures reduce the internal the water vapor concentrations in CAM plants. This in turn is the key to the Water-Use Efficiency that we will discuss in the next section. But first let us bring both CO_2 uptake and transpiration into the picture for CAM plants.

To portray the daily (24-hour) gas exchange patterns for CAM plants on a graph, the abscissa (*x*-axis) usually begins near noon, when net CO_2 exchange (Fig. 2-3) and water vapor loss by such species are minimal. Figure 2-4 shows that the maximal net CO_2 uptake rate for *Agave deserti*, commonly known as the "desert agave," is about 6 μmol m^{-2} s^{-1}, which is a relatively low maximal net CO_2 uptake rate among plants. Specifically, *A. deserti* sacrifices high net CO_2 uptake capabilities by having stomata that markedly restrict gas exchange, resulting in low rates of water loss by transpiration. This underscores the over-riding importance of water conservation by this species native to deserts in arid and semi-arid regions.

Besides being the "desert agave," *A. deserti* is also commonly referred to as a "century plant." This is because its relatively low CO_2 uptake ability translates into a slow growth rate. Actually, it flowers once and then dies after about 50 to 60 years (Nobel, 1988). Hence the moniker "century" is a bit exaggerated (the same naming situation applies to *Agave americana*, which actually has a much shorter lifetime). In any case, it is a much poorer producer of biomass than is *Agave tequilana* (the "blue agave" or "agave azul") of tequila fame (Chapter 1), which we consider next.

In contrast to *A. deserti*, *A tequilana* in cultivation flowers after only 6 to 9 years. Such flowering is an anathema to the tequila industry, as all of the carbohydrate "goodies" that could have been converted to tequila are then "wasted" to produce the inflorescence. Hence plants are "castrated" (also mentioned in Chapter 1) by whacking off a young inflorescence, which grows (extends) from its top. This arrests the growth of the inflorescence and stops the export of sugars from the leaves and more importantly from the stem or piña (see Chapter 1) to the inflorescence, which for this species can become 6 meters (20 feet) tall. In any case, *A. tequilana* (Fig. 2-4B) has a much higher maximum rate of net CO_2 uptake, 15 μmol m^{-2} s^{-1}, than does *A. deserti* (Fig. 2-4A).

The champion CAM plant with respect to net CO_2 uptake rate is *Opuntia ficus-indica*, whose uses for vegetables, forage, fodder, cochineal insects, and especially fruits are legendary (Chapter 1). It can achieve a maximal net CO_2 uptake rate of about 25 μmol m^{-2} s^{-1} under ideal growing conditions (Fig. 2-4C). Such a rate surpasses the maximal CO_2 uptake rates for a large proportion of C_3 and C_4 species (see Fig. 2-3 for average values for these plants).

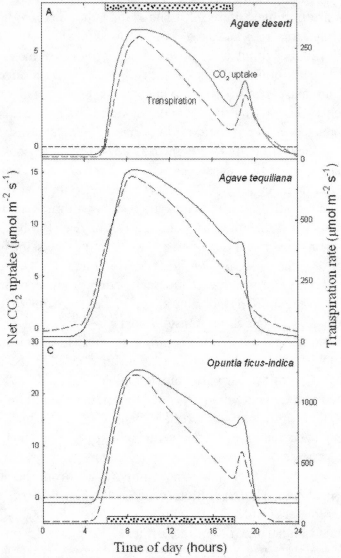

Figure 2-4. Daily patterns of net CO_2 exchange rates (solid lines) and transpiration rates (dashed lines) for three CAM species: (A) *Agave deserti* in the northwestern Sonoran Desert, California (Nobel, 1988, 2009); (B) *Agave tequilana* in Jalisco, Mexico (Nobel and Valenzuela, 1987; Nobel et al., 1998; Pimienta et al., 2001); and (C) *Opuntia ficus-indica* in Jalisco, Mexico and in the laboratory (Pimienta et al., 2000; Nobel, 2001). Measurements were made under essentially ideal light conditions (clear days), optimal temperatures [minimum nighttime temperatures

of 10°C (50°F), average nighttime temperatures of 14°C (57°F)], and wet soil. As for Figure 2-3, the stippled region indicates the night. Note the different scales used on the ordinate (y-axis) in the three panels.

Now let us consider the daily timing of transpiration and its consequences for these three CAM species. We know that transpiration is inevitable, as even when the stomata are closed, water is lost by cuticular transpiration. But the stomata must be open at night for substantial nocturnal CO_2 uptake by CAM plants to occur (Figs. 2-3 and 2-4). More important, the rates of nocturnal transpiration by CAM plants are much lower than are the transpiration rates during the daytime by C_3 and C_4 plants for the same degree of stomatal opening.

Figure 2-4 indicates maximal transpiration rates of 270 μmol m^{-2} s^{-1} for *A. deserti*, 670 μmol m^{-2} s^{-1} for *A. tequilana*, and 1200 μmol m^{-2} s^{-1} for *O. ficus-indica*. (Transpiration rates are usually expressed in millimoles m^{-2} s^{-1} but here we are using micromoles m^{-2} s^{-1}, as for photosynthesis. This will facilitate comparisons between these two gas exchange processes in the next section.) Maximal daytime transpiration rates by C_3 and C_4 plants under the same environmental conditions would average four- to eight-fold higher. The reason for the higher transpiration rates of C_3 and C_4 plants during the daytime and the consequences are discussed in the next section. We note here that for all three CAM species in Figure 2-4, transpiration is predominantly at night. This water loss causes a small but measurable decrease in thickness of about 0.2% for the leaves of the agaves and the cladodes of the platyopuntia. A similar decrease in diameter occurs for barrel and columnar cacti at night when they have substantial rates of gas exchange.

For all of the three CAM species whose gas exchange patterns are presented in Figure 2-4, we notice a curious bump up of net CO_2 uptake and transpiration at dawn. This is not an artifact but rather an important physiological response of the stomata and the enzymes to the environmental conditions prevailing at the end of the night. Anyone who wakes at dawn and goes outside notices how cool or cold it is then. Soon the sun comes up and warms everything. CAM plants also "recognize" the upcoming warming trend and so sneak over to act like C_3 plants for a short time under the cooler conditions prevailing at dawn.

Specifically, the binding of atmospheric CO_2 shifts from PEPCase at night, as is typical for CAM plants (Fig. 2-1C), to Rubisco at dawn, which is typical for C_3 plants during the daytime (Fig. 2-1A). To facilitate the entry of CO_2 at dawn, the stomata open more, which accounts for the bump up in the transpiration rate then (Fig. 2-4). The increased transpiration occurs when the potential water loss rate is minimal because of the low temperatures at that time. But more about that in the next section, which is the most important one for understanding the special water-conserving physiological behavior of agaves, cacti, and other CAM plants.

Water-Use Efficiency, CAM, and Drying Clothes

We now have within our grasp an understanding of the great advantage of CAM plants. So now let us consider its "nuts and bolts." This is quantified by the *Water-Use Efficiency*, or WUE for short. This can be simply stated as follows (Nobel, 2009):

$$\text{WUE} = \frac{\text{amount of } CO_2 \text{ fixed by photosynthesis}}{\text{amount of water lost by transpiration}} \quad (2.3)$$

The WUE indicates the benefit (CO_2 uptake) per unit cost (H_2O loss). Its reciprocal, which is called the *Transpiration Ratio* (TR), is often more easily appreciated and indicates the water lost per CO_2 fixed. Clearly it is preferable (1) to lose as little water as possible per unit amount of CO_2 taken up (low TR), and (2) to take up as much CO_2 as possible per unit amount of water lost (high WUE). Remarkably, this is exactly what stomata control and do! Nobody knows all of the cellular processes and feedback mechanisms involved in this balancing act. Nevertheless, these tiny pores in leaves and stems have the "smarts" needed for economically using the water taken up from the soil for the uptake of carbon dioxide from the atmosphere, which sustains life on Earth.

Ecologically speaking, the daily or seasonal Water-Use Efficiency and Transpiration Ratio are more important, but let us first consider gas exchange on an instantaneous basis. For the three CAM species in Figure 2-4, net CO_2 uptake is maximal early in the night at about

21:00 hours (9:00 pm). Transpiration is also approximately maximal then, indicating that the stomata are open. The TR is relatively easy to calculate here, because both the left ordinate (the CO_2 uptake rate) and the right ordinate (the transpiration rate) in Figure 2-4 are in the same units (namely, $\mu mol\ m^{-2}\ s^{-1}$).

For *Agave deserti* (Fig. 2-4A), the Transpiration Ratio at 21:00 hours is 270 $\mu mol\ H_2O$ lost $m^{-2}\ s^{-1}$ divided by 6 $\mu mol\ CO_2$ fixed $m^{-2}\ s^{-1}$, or 45 H_2O lost per CO_2 fixed. For *Agave tequilana* (Fig. 2-4B), the TR is then 43 H_2O lost per CO_2 fixed, and for *Opuntia ficus-indica* (Fig. 2-4C), it is 48 H_2O lost per CO_2 fixed. Thus, 40 to 50 water molecules are lost for each molecule of carbon dioxide incorporated into a carbohydrate. This may seem like a lot of water lost. But if we had done the same calculation for C_3 or C_4 plants during the daytime, when their maximal rates of net CO_2 uptake take place, we would find that generally 200 to 400 molecules of water are lost per molecule of carbon dioxide fixed in photosynthesis. What gives? It is the daytime versus nighttime temperature at these times for maximal gas exchange, as we have already hinted. Indeed, the water vapor concentration in the leaves increases dramatically as the temperature increases. Thus, more water loss occurs at higher temperatures for the same degree of stomatal opening.

Gases move into and out of leaves by diffusion. Net movement by diffusion is from regions where the molecules are more concentrated to regions where they are less concentrated (Fick's first law of diffusion; Nobel, 2009). Some molecules move in the "wrong" direction, so we must consider all of the molecules on a statistical basis. Imagine the end of a football game where everybody is piling out of the stadium (high concentration region) to the parking lot or to the bus stop (low concentration region). A few people will be going in the wrong direction back to the stadium to retrieve their forgotten coats, binoculars, or cell phones, but the net flow is clearly outward. Likewise, CO_2 has a net diffusion into a C_3 or C_4 plant during the daytime, because Rubisco and PEPCase have reduced the internal CO_2 concentration compared with the atmospheric CO_2 level.

Similarly, molecules of water vapor diffuse out of a leaf, where their concentration is high, to the atmosphere, where their concentration is usually low. Indeed, the relative humidity (Equation

2.2) in the air between cells within a leaf, the *intercellular air spaces*, is nearly 100%, whereas the relative humidity is generally much lower in the atmosphere. And the concentration of water vapor at 100% relative humidity (the saturation value) increases approximately exponentially with temperature.

Let us next be specific about the relationship between temperature and the saturated water vapor content of air. Air at 0°C (32°F) saturated with water vapor contains 0.27 mole of water vapor per cubic meter of air. At 10°C (50°F), this concentration becomes 0.52 mol m^{-3}. At 20°C (68°F), 30°C (86°F), and 40°C (104°F), the content of water vapor at saturation becomes 0.96, 1.69, and 2.84 mol m^{-3}. That is, air can hold much more water vapor at higher temperatures. Because of this, the relative humidity (Equation 2.2) is generally lower during the daytime than at night when the air temperatures are lower (Nobel, 2009). [Admittedly, presenting temperatures on both the Celsius and the Fahrenheit scales slows down the reading. However, our feeling for temperatures is generally very personal. So please choose your preferred scale and ignore the other.]

The exponential increase in the saturated water vapor content of air with increasing temperature is crucial for the water-conserving ability of CAM plants. In particular, the intercellular air spaces in a leaf (or a stem) are basically saturated with water vapor. For our initial calculation, let us assume that the air surrounding such a leaf contains 0.40 mol water vapor m^{-3}, as is typical of the Sonoran Desert during the winter. The water vapor concentration of this air does not vary much over the course of a day unless a change in the weather occurs, such as a weather front bringing in rainfall.

As just indicated, transpiration is driven by the higher water vapor concentration in a leaf or a stem versus the lower water vapor concentration in the air. Thus, water vapor tends to diffuse out of a leaf. For a leaf at 20°C (68°F) with 100% relative humidity inside (0.96 mol water vapor m^{-3}), the leaf-to-air water vapor concentration difference is 0.96 mol water vapor m^{-3} in the leaf minus 0.40 mol water vapor m^{-3} in the surrounding air or 0.56 mol water vapor m^{-3}. Next let us calculate the leaf-to-air water vapor concentration difference, which is the driving force for transpiration, at a lower leaf temperature. For instance, at 10°C (50°F), these water vapor concentration values would be 0.52 mol water vapor m^{-3} at 100% relative humidity inside a leaf minus 0.40 mol water vapor m^{-3} in the outside air, or 0.12 mol m^{-3} as the leaf-to-air water vapor concentration difference. For the same degree of stomatal opening, the transpiration rate would then be $(0.56$ mol $m^{-3})/(0.12$ mol $m^{-3})$ or five times higher at 20°C than at 10°C. Clearly, transpiration is much greater at the higher temperature.

Let us next compare leaves at 30°C (86°F) during the daytime with 20°C at night and assume a realistically higher air water vapor concentration of 0.80 mol m^{-3} in the surrounding air under these warmer conditions. The leaf-to-air water vapor concentration drops then are $1.69 - 0.80$ or 0.89 mol m^{-3} at 30°C versus $0.96 - 0.80$ or 0.16 mol m^{-3} at 20°C. Thus transpiration would be $(0.89$ mol $m^{-3})/(0.16$ mol $m^{-3})$ or six-fold higher at the higher daytime temperature of 30°C than at the lower nighttime temperature of 20°C, if the stomatal opening were the same in the two cases. Clearly, the force for diffusion of water vapor out of the leaf is again much lower for the lower nighttime temperature. This underlies the advantage of CAM in limiting water loss from the plants. Namely, the concentration of water vapor in the intercellular air spaces in the leaves of CAM plants is much lower at night when the stomata are open than during the daytime when their stomata are closed. Consequently, the force leading to transpiration is much lower at night.

Before returning to the Water-Use Efficiency of C_3, C_4, and CAM plants, let us consider a simple but basically correct analogy—the drying of clothes outdoors. We all know that clothes dry faster outside during the daytime than at night. Wet clothes, such as a shirt, a bathing suit, or a large beach towel, basically have a saturated water vapor concentration at their surfaces, similar to that occurring inside a leaf. Water vapor thus diffuses away from the clothes to the atmosphere, where the water vapor concentration is much lower. At night, the lower temperatures mean that the water vapor concentration drop from the wet clothes to the atmosphere is much less, so the rate of diffusion of water vapor away from the surface of the clothes is less. Hence the drying of the clothes is much slower at night than during the daytime. This is the crucial principle behind the advantages of the nocturnal opening of stomata by CAM plants, as we have already suggested.

Now for some actual values of WUE (Equation 2.3) typical of the three photosynthetic pathways (Nobel, 2009). Of course, all values depend on the environmental conditions, which affect both transpiration and photosynthesis. We first note that the Transpiration Ratio can be higher and the Water-Use Efficiency correspondingly lower when considered over 24-hour periods than for the instantaneous maximal values with the stomata open. In particular, such 24-hour periods include times with both water loss and net CO_2 loss, such as at night for C_3 and C_4 plants and during the daytime for CAM plants (Figs. 2-3 and 2-4), which reduces the Water-Use Efficiency. On the other hand, partial stomatal closure reduces transpiration more than it reduces net CO_2 uptake, the latter process involving many more steps (Nobel, 2009). This occurs as the soil begins to dry, which reduces net CO_2 uptake but raises the WUE (Equation 2.3).

The WUE generally averages 0.0013 to 0.0050 CO_2 fixed per H_2O lost for C_3 plants (TR of 200 to 800 H_2O's per CO_2). C_4 plants tend to have higher photosynthetic rates than C_3 plants (Fig. 2-3) and also they have less stomatal opening, which lowers transpiration with only a modest effect on net CO_2 uptake. Thus, WUE generally ranges from 0.0025 to 0.010 CO_2 per H_2O for C_4 plants (TR of 100 to 400 H_2O's per CO_2).

The Water-Use Efficiency is often more variable for CAM plants. In part this is because CAM plants can have both nighttime and daytime uptake of CO_2 (Fig. 2-4), whereas C_3 and C_4 plants can take up CO_2 only during the daytime (Fig. 2-3). In any case, the WUE for CAM plants is much higher than for plants using the other two photosynthetic pathways. In particular, their WUE is generally 0.013 to 0.040 CO_2 per H_2O (TR of 25 to 80 H_2O's per CO_2). This is why CAM plants are so well suited to arid and semi-arid habitats and relatively dry microhabitats. Taking the mean values of the ranges just presented, the WUE averages two-fold higher for C_4 plants than for C_3 plants and eight-fold higher for CAM plants than for C_3 plants. The crucial implications, in terms of global climate change, for the much higher Water-Use Efficiencies of CAM plants will be considered in Chapters 5, 6, and 7. But the bottom line is that

CAM plants can use water much more efficiently with regard to CO_2 uptake and productivity than do C_3 and C_4 plants.

Evolution and Family Distribution of CAM

So far we have sung the praises of the water-conserving aspects of CAM. Let us next consider this photosynthetic pathway from an evolutionary perspective. Based on the presence of CAM in many unrelated taxa, it probably has evolved independently five or more times among vascular plants (Winter and Smith, 1996). CAM occurs in the monocot family Agavaceae and the unrelated dicot family Cactaceae (Preface, Chapter 1), as we have noted. It occurs in the Polypodiaceae and the Vittariaceae, two families of ferns. Indeed, CAM plants have been successful in a wide range of habitats and microhabitats.

We have already mentioned that most CAM plants are epiphytes and hemiepiphytes, whose microhabitats can be very dry a few days after a rainfall. One of the most curious habitats for CAM plants is below the surface in fresh-water lakes, where CO_2 levels are greatly reduced during the daytime by the photosynthesis of various water plants. At night, these organisms plus animals release CO_2, so CAM plants that "work the night shift" are presented with an opportunity—high nocturnal levels of CO_2. *Isoetes howellii* (family Isoetaceae), a submerged aquatic CAM plant, is a case in point. This species and others in the genus *Isoetes*, which occur on five continents, show increases in malate levels and hence increases in acidity during the night. Family Crassulaceae from which CAM got its name has the genus *Crassula*. Some of its 300 species are also aquatic like *I. howellii*, and essentially all of these exhibit nocturnal increases in acidity and net CO_2 uptake then.

Terrestrial plants most likely utilized C_3 photosynthesis 420 million years ago (Silurian epoch; Raven and Spicer, 1996). Atmospheric CO_2 levels then were probably about 10,000 ppm and O_2 levels were slightly less than the current value of 21%. Under such conditions, photorespiration would be negligible and plenty of carbon dioxide was available to plants. The stomata would not need to open very much to let sufficient CO_2 into a leaf, and so transpiration would be relatively low. Hence, the advantage of CAM in terms of higher Water-Use Efficiency would not be very important.

By the late Carboniferous epoch just over 300 million years ago, atmospheric CO_2 levels were similar to current values and the water-conserving aspect of CAM would be useful. The next time that the atmospheric CO_2 levels were again so low was in the Pleistocene epoch beginning about 2 million years ago. In any case, CAM plants could have evolved over a wide geologic time range. Although we cannot pinpoint the date for the origin of CAM, we can speculate that it evolved as plants colonized dry habitats or dry microhabitats (epiphytes, hemiepiphytes). We also note that the Carboniferous epoch (300 to 360 million years ago) has a connection to global climate change, as its name derives from the Latin *carbo*, meaning coal. Indeed, much of the worldwide coal beds are derived from plants living in that epoch.

Let us return to the taxonomic distribution and occurrence of CAM. It has been demonstrated in 40 families out of the total of approximately 500 families of vascular plants. Besides the fern allies of the Isoetaceae, tropical epiphytic ferns (two families), and the gymnosperm *Welwitschia mirabilis* (family Welwitschiaeae), plants in 10 families of monocots exhibit CAM, as do plants in 26 families of dicots. The greatest number of species of CAM plants occurs in the monocot family Orchidaceae, probably about 80% of its approximately 25,000 species. Thus this largest family of the angiosperms may contain 20,000 CAM species.

The family Agavaceae as viewed here has nearly 300 species (Gentry, 1982; Good-Avila et al., 2006), although some taxonomists include more species. In any case, its largest genus is *Agave* with approximately 140 species (range in the literature of 113 to 166 species), apparently all of which are leaf succulents exhibiting CAM. Probably 99% of the just over 1,600 species of family Cactaceae are capable of CAM. The cacti that are C_3 at maturity are leafy ones, most of which occur in the relatively primitive genus *Pereskia* (Gibson and Nobel, 1986), whose large leaves are not unlike those of ivy. Another 1% of the Cactaceae are even more curious, having C_3 leaves (at least most of the time) and CAM stems.

Over the years, many "cactologists" have contributed to our understanding of the origin of the Cactaceae. One consensus is that this occurred about 70 million years ago (end of Cretaceous epoch) in the Caribbean and northern South America (Wallace and Gibson, 2002). The Cretaceous epoch was 65 to 145 million years ago, when the supercontinent Pangaea broke up into our present continents. With respect to global climate, extensive beds of calcium carbonate, used in cement manufacture, were laid down during this period, especially in present-day Europe. Indeed, the word Cretaceous is derived from the Latin *creta*, meaning chalk. Rocks were also formed during this epoch that are important sources for oil and gas. In addition, the great *Tyrannosaurus rex*, a dinosaur for all seasons, lived in the Cretaceous epoch.

Other estimates for the origin of the Cactaceae place it more recently at about 30 to 40 million years ago, probably in central South America (Metzing and Kiesling, 2008). The Agavaceae evolved somewhat later than the Cactaceae, perhaps 30 million years ago. Within this family, the genus *Agave* is of more recent origin, probably 10 million years ago in present-day Mexico (Good-Avila et al., 2006). Even though it is crucial to know how we got here, our focus in this book is on the future for agaves and cacti.

Before moving on, let us comment on a feature that relates to the timing of the evolution of C_3 versus CAM pathways. C_3 plants cannot perform CAM, whereas CAM plants can and do perform C_3 metabolism (Nobel, 1988; Raven and Spicer, 1996). This is part of the reason for stating that C_3 is more primitive than CAM. Interestingly, nearly all of the CAM species from arid and semi-arid regions that have been studied act as C_3 plants as seedlings, when water from rainfall is usually plentiful. *Agave deserti* raised from seed in the laboratory under wet conditions has predominantly daytime CO_2 uptake for approximately 6 months after germination. For *Opuntia ficus-indica*, young cladodes develop in the spring on underlying "mother" cladodes, which hence are providing the water for the growth of the young cladodes. At 2 weeks of age, the cladodes have predominantly daytime stomatal opening and CO_2 uptake, whereas at 4 weeks of age, the gas exchange pattern is predominantly CAM.

A revealing mistake with regard to photosynthetic pathway occurred when we were simultaneously studying desert ferns and *A. deserti* in my laboratory. A volunteer worker got the watering instructions backwards, so he watered the agaves every other day and withheld water from the fern for 3 months. The ferns shriveled up, whereas the mature 20-year-old agaves shifted from net CO_2 uptake only at night to 97% of the total daily net CO_2 uptake occurring during the daytime (Hartsock and Nobel, 1976). The accidental mix-up showed that this CAM species could perform almost indistinguishably from a C_3 plant if supplied with plenty of water. Again, water conservation is the trademark of the advantage of CAM plants.

3

Special Drought and Temperature Tolerances

As indicated in the Preface and as will be considered in detail in the next chapter, global climate change will involve increasing temperatures and variations in rainfall patterns. Thus we need to know the responses of agaves and cacti to these two environmental parameters to make predictions both in time (the future) and in space (specific geographic regions). This chapter will focus on survival—what the plants can tolerate before dying. Chapter 5 will address the effects of such environmental parameters on net CO_2 uptake and hence productivity. But to produce, plants must be alive, so we first explore the tolerances of agaves and cacti to drought and then their tolerances to extreme temperatures, both hot and cold.

Drought—From Deserts to the Tropics

Water limits agricultural productivity worldwide. Therefore, irrigation, which currently consumes 70% of the freshwater used by humans, is a common practice that is increasingly expensive. Sometimes not enough water is available for all who wish to use it. On the other hand, dryland farming and natural ecosystems rely on rainfall. In any case, the period when plants cannot take up water from the soil, either because of insufficient irrigation or too little rainfall, is referred to as a *drought*.

This simple enough definition of drought requires some explanation to interpret it in terms of the quantites involved. We begin by noting that water tends to flow from higher to lower hydrostatic pressure, where *hydrostatic pressure* is the force per unit area in the liquid. Consider the plumbing systems in our houses or in our bodies. When we open a faucet, water flows from the higher hydrostatic pressure in the pipes to the lower hydrostatic pressure region in a sink. Likewise, blood, which is mainly water, flows from the contracting left ventricle of our heart out to the lower hydrostatic pressures in the aorta and then the arteries. For plants, water tends to flow from a higher hydrostatic pressure in a wet soil to a lower hydrostatic pressure in the roots.

Water movement also involves the effect of salts and other molecules in the solutions. It is the same principle as for the movement of water vapor in transpiration discussed in the previous chapter. There we noted that water vapor diffuses from a leaf, where its concentration is high, out to the surrounding air, where its concentration is nearly always lower. Here we note that liquid water moves from where it has a higher concentration (meaning a lower solute concentration) to regions where water has a lower concentration (meaning a higher solute concentration).

Water hence flows from a wet soil, which has few dissolved solute molecules per unit volume of the soil water, to the cells of a root. The cells can have many solute molecules per unit volume and hence a lower water concentration, because the solute molecules displace some of the water molecules (Nobel, 2009). *Osmotic pressure* is proportional to the number of solute molecules per unit volume—the more solute molecules, the higher is the osmotic pressure. Thus water tends to flow toward regions of higher osmotic pressure (lower water concentration), a process called *osmosis*.

We can predict the direction for water flow based on the hydrostatic pressure, where water tends to flow toward lower values, and the osmotic pressure, where water tends to flow toward higher values. Fortunately, these two effects are algebraically additive. Water thus enters a plant based on some combination of flowing toward lower hydrostatic pressure and higher osmotic pressure. When the soil dries, its hydrostatic pressure decreases, as less water is then available among the soil particles. Also as a soil dries, its osmotic pressure increases (the same number of solute molecules dissolved in less water, so the solute concentration increases). Therefore, water has less tendency to enter a plant as the soil dries, as we would expect. Drought begins when the soil dries to the point where the hydrostatic and the osmotic pressures in the soil and in the plant balance each other, so no net water uptake occurs.

Practically speaking, agaves and cacti are like bags of water. Thus they tend to have low internal osmotic pressures and tend to take up water only from very wet soils. But they also tend to have shallow roots, so they can respond to even light rainfalls. For instance, the mean root depths of the desert agave *Agave deserti* and the barrel cactus *Ferocactus acanthodes* in the northwestern Sonoran Desert in California are only 10 centimeters (4 inches; Nobel, 1988). For these species in dry sandy soil, about 7 millimeters (just over ¼ inch) of rainfall is sufficient to tip the balance from drought to water uptake. We have already mentioned that most CAM plants are epiphytes or hemiepiphytes, including more than 100 species of cacti growing in moist tropical regions (Chapter 2). For them, water uptake can occur after even a light morning shower, as the water trickles down a trunk and accumulates in a tree crotch where their roots can occur. But no rain, no gain!

Consequences of Anatomy/Morphology

We have just discussed the quantities involved in the uptake of water by a plant. The next logical question to ask is "Can such water be stored in plants as a preventative measure to avoid the consequences of drought?" Plant cells were discussed in Chapter 2, where those using the CAM pathway had an increase in acids during the night (Fig. 2-1; the "A" of CAM stands for "Acid"). This nocturnal accumulation of acids occurs in a large cellular region referred to as the *central vacuole*, which can occupy 90% of the volume of a *chlorenchyma* (chlorophyll-containing and hence photosynthetic) cell of a CAM plant. This central vacuole is part of the *anatomy* of CAM plant cells, and its large size favors the possibility for cellular water storage. More apparent is the *morphology* of CAM plants, which are characterized by succulent (thick and juicy) leaves and succulent stems (Fig. 1-1). Large amounts of water can be stored in the leaves and the stems of such plants—again, they are "water bags."

The water stored in the cells and the organs of agaves and cacti can be called upon to maintain them during drought. The typical photosynthetic organs of C_3 and C_4 plants—their relatively thin leaves—have no such water reservoirs, and thus these plants are more vulnerable to water loss during drought. In other words, the water storage of the massive leaves of agaves and the massive stems of cacti causes them to have a large water reservoir available per unit surface area across which water is lost during transpiration. Moreover, during the initial phases of drought, these CAM plants tend to shift water from the whitish internal water-storage cells to the greenish external cells (chlorenchyma) where photosynthesis takes place. This cellular water altruism is another crucial feature of CAM plants.

Typically, C_3 and C_4 plants suffer irreparable damage once they lose 30% of their water content. However, many cacti can survive an 80 to 90% loss of their hydrated water content and still survive, although in a very emaciated condition. Such cacti can swell up as they become rehydrated in a matter of days to a few weeks following heavy rainfall—such a revival allows them to be prepared for the next drought. Thus agaves and cacti take advantage of three abilities: (1) the ability to store a lot of water, (2) the ability to shift water around among cells to keep crucial metabolism active, and (3) the ability to tolerate extreme cellular dehydration.

Besides storing a lot of water, agaves and cacti can also restrict the loss of such stored water across their surfaces. In Chapter 2, we indicated that transpiration can be greatly reduced by closing the stomata tightly. Greatly reduced but not eliminated, as some water moves across the cuticle, the *cuticular transpiration*. The cuticle is the waxy covering on the surfaces of leaves and green stems and contains *cutin*, as already mentioned. Cutin is a mixture of fatty substances that is fairly inert chemically and a good water-proofing agent to boot.

The cuticles of agaves and cacti can be 5 to 40 micrometers (μm; 0.2 to 1.6 thousandths of an inch) thick. On the other hand, those on the leaves of C_3 and C_4 plants generally average slightly less than 1 μm in thickness (Nobel, 1994). The extra thickness of the cuticles for these CAM plants provides a much greater barrier for water loss. Hence a very low rate occurs for their cuticular transpiration. The cuticles are so thick and strong on leaves of *Agave atrovirens* and *Agave salmiana* that in Mexico they are stripped off and used to wrap tortilla sandwiches, meat, and even some tasty casseroles.

Consequences of CAM

But the anatomical and the morphological features of CAM plants are not their only defense against droughts. They also have a physiological one, the use of Crassulacean Acid Metabolism (Chapter 2). Although the calculations in the last chapter about Water-Use Efficiency may have seemed a bit tedious, they indicate that water is lost relatively slowly by CAM plants compared with C_3 and C_4 plants. In particular, the rate of transpiration is only about 15 to 20% as high for a CAM plant with nocturnal stomatal opening compared with the same degree of stomatal opening for C_3 or C_4 plants, whose stomata open during the daytime. We all know that clothes dry slowly at night. Limited transpiration is crucial for the increased future uses of agaves and cacti as rainfall patterns shift and water for irrigation becomes increasingly expensive as well as problematic.

Another consequence of the lower amount of water used per CO_2 fixed by CAM plants compared with C_3 and C_4 plants is less need for a massive root system for water uptake. For many non-CAM perennials, the roots represent 30 to 50% of the total plant biomass; but for agaves and cacti, the root system is much smaller, generally only 8 to 12% of the total plant biomass (Nobel, 1988). The lower proportion of valuable resources diverted to roots leads to a greater proportion available for the shoots. This is important for supporting photosynthesis by the shoots. Also, the shoots are readily harvested, so extra resources available to them contribute to the high productivity possible for certain agaves and cacti.

Incredible Survival Records

The highly respected authority for records, formerly *The Guiness Book of World Records* and now titled *Guinness World Records*, relates many incredible feats. But apparently no records are listed for the length of plant survival during drought, which we consider next.

My first introduction to this topic was with the platyopuntia *Opuntia basilaris* in the Sonoran Desert of California. Its stems could be removed from the ground and yet the plants would survive for three years (Ting and Szarek, 1975). My second contact with long droughts was for the small barrel cactus *Copiapoa cinerea* in northern Chile. It had not rained for six years, but this cactus was still alive, although not looking so pretty (Nobel, 1988). In the laboratory various species of barrel cacti in the genus *Ferocactus* can be exposed to drought for more than one year; when they are rewatered, the stems become greener and net CO_2 uptake resumes during the first week. These incredible survival times are testimony to the ability of their stems to store and to ration water, and especially to the extremely low rates of cuticular transpiration, which is a consequence of the thick and relatively water-impervious cuticles.

Many of you have had personal experiences with household and garden cacti that you have on purpose or by accident failed to water for many months. Upon rewatering, the plants recovered and are obvious champions for surviving drought. We will return to this theme of the ability of CAM plants in general, and agaves and cacti in particular, to survive drought after we view the changing rainfall patterns expected in the future as part of global climate change, which is discussed in the next chapter. In any case, it is relatively hard to kill agaves and cacti by drought, although their net CO_2 uptake can be reduced to zero by drought in a month or so, as is discussed in Chapter 5.

High Temperatures—You Can Fry an Egg

Some agaves and cacti live in incredibly hot places. To describe such regions, air temperatures in United States weather stations are typically measured at approximately 1.5 meters (5 feet) above the ground level. In the northwestern Sonoran Desert in Arizona and southeastern California, where much of the ecophysiological research on agaves and cacti has taken place (Nobel, 1988), such air temperatures annually reach about 47°C (117°F). These measurements actually do not do justice to the temperatures that plants must tolerate closer to the ground. Indeed, when the official air temperature is 47°C, the soil surface and the air immediately adjacent to it can reach 70°C (158°F). It is so hot that you can literally fry an egg placed on the ground or on the hood of an automobile.

Although not agaves or cacti, species of *Lithops*, a genus native to Namibia and South Africa whose plants grow right at the soil surface, can teach us much about high-temperature tolerance. They are called "stone plants," because they are cryptically disguised as pebbles in the deserts where they occur. These small plants have essentially the same temperature as the soil surface, which routinely exceeds 60°C (140°F) in the summer there (Nobel, 1989). However, most vascular plants die at 50 to 55°C (122 to 131°F), so the high-temperature tolerance of *Lithops* is special. We next explore how to measure the high temperatures tolerated by plants, such as the temperatures experienced by *Ferocactus acanthodes* and other succulent cactus species in the northwestern Sonoran Desert, especially near the soil surface.

Cells—Measuring Tolerances

When we began investigating the temperature extremes that can be tolerated by agaves and cacti, we needed a method to tell when a cell was dead. This sounds easy, but death is hard to define at a cellular level and also not so simple for a whole plant. For instance, various columnar cacti in Arizona, such as *Carnegiea gigantea* (saguaro) and *Stenocereus thurberi* (organ pipe cactus), were exposed to unusually cold and prolonged freezing episodes in the 1930s. Damage was evident in months but some of the cacti did not die for three years. Waiting three years to see the results of a specific experimental treatment is not practical in our hurry-up world.

Thus, we used various means to determine whether a high-temperature treatment (1) disrupted membranes (which control what goes into and comes out of a cell or organelle), (2) denatured proteins (meaning making them non-functional), or (3) damaged mitochondria (the cellular organelles responsible for producing ATP, an important energy currency that we will discuss in Chapter 6). We settled on a relatively simple technique visually apparent using a light microscope—the intracellular accumulation of a "vital" dye, which occurs for living cells only (hence the word "vital").

The dye is called *neutral red*, with the official chemical name of 3-amino-7-dimethyl-amino-2-methylphenazine hydrochloride, which does not easily roll off the tongue! Neutral red accumulates in the central vacuoles of living cells, which are especially large for CAM plants. Under a light microscope, such living cells appear pink to red in the presence of neutral red. When a treatment kills the cells, none of this dye is taken up into their central vacuoles, and so the cells then appear clear to faint orange. Admittedly, some judgment is necessary to distinguish light orange from pinkish and some skill is needed not to be confused by colors in overlying or underlying cells. But an undergraduate student can master the technique of deciding whether a cell is alive or dead in a few hours at a microscope.

Armed with a set of light microscopes and temperature measuring paraphernalia, in the late 1970s I took an undergraduate class on an ecological field trip lasting two days under primitive conditions. The purpose of the trip was to make some of the first measurements of the temperature tolerance of agaves and cacti under field conditions. I told the students that the "women's hill" is to the north and that the "men's hill" is to the south, but many were obviously embarrassed. When we left, we stopped at the nearest gas station. You should have seen the expression on the attendant's face as nearly all of the students piled out of the university research vehicles and dashed for the restrooms!

The class soon had collected data indicating that these desert succulents can have phenomenal tolerances of high temperatures. These studies were followed up in the laboratory (Fig. 3-1), where conditions are easier to control (and bathrooms are nearby). Over the years, the cellular uptake of neutral red was correlated with the death throes experienced by plants exposed to extreme temperatures. The bottom line is that when none of its cells take up the vital dye, the plant eventually dies. Because of time and labor limitations, "none" is often hard to determine in the laboratory. Thus, the more easily determined treatment temperature that reduces dye uptake by 50% is found; this temperature is based on microscopic measurements that are presented graphically.

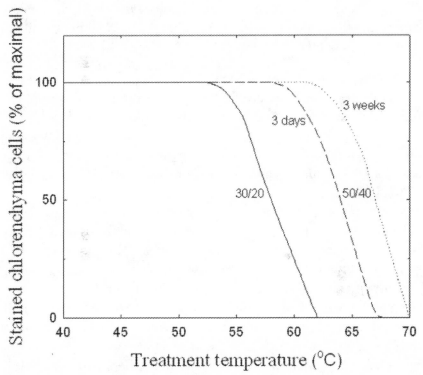

Figure 3-1. Decrease in the uptake of neutral red into the chlorenchyma cells of *Ferocactus acanthodes* as the treatment temperature is raised. Plants were maintained in environmental chambers with moderate day/night air temperatures of 30°C/20°C (86°F/68°F; solid line). Plants were then transferred to high day/ night air temperatures of 50°C/40°C (122°F/104°F) for 3 days (dashed line) or for 3 weeks (dotted line). After placing stem pieces in an oven for 1 hour at the indicated treatment temperature, thin slices about three cell layers thick were exposed to the vital dye neutral red for 30 minutes at 25°C (77°F). After washing away excess dye, the slices were observed using a phase-contrast light microscope to see whether neutral red appeared in the central vacuoles of the chlorenchyma

cells. In particular, a reddish color (stained) in the cells indicated that they were living, whereas a pale color (unstained) indicated the disruption of membranes that accompanies cell death (Nobel, 1988; Nobel and Zutta, 2008).

On average, plants die when placed under a 2.8°C (5.0°F) colder low-temperature treatment than the one leading to the readily measured halving of the cellular ability to take up neutral red (Nobel, 1988; Nobel and Zutta, 2008). And plants die at a 4.0°C (7.2°F) hotter high-temperature treatment than the one causing a 50% decrease in stain uptake. The simple test of ascertaining the treatment temperature that halves stain uptake for a small plant sample therefore allows the determination of a thermal limit without killing whole plants.

Acclimation—A Key to Survival

Let us next check the quantitative details for the cellular uptake of a vital stain as presented in Figure 3-1, which shows data for the barrel cactus *Ferocactus acanthodes* growing under two day/night air temperature regimes. Nothing much happens to the stain uptake ability as the treatment temperature is raised from 40°C (104°F) to 55°C (131°F), a relatively high temperature that certainly would be very uncomfortable for us. The treatment time was standardized at 1 hour, a typical time that an agave leaf or a cactus stem might experience an extreme temperature in the field (Nobel, 1988, 1994). As the treatment temperature is raised further (see legend to Fig. 3-1 for experimental details), plants growing under the lower (more moderate) day/night temperatures, 30°C/20°C (86°F/68°F), begin to show a decrease in the cellular ability to accumulate neutral red. The decrease is gradual at first and then accelerates as the treatment is increased further, making it easy to determine where a 50% reduction in stain uptake occurs.

In particular, the halving of the ability to take up neutral red occurs at 58°C (136°F) for *F. acanthodes* growing under moderate temperatures (Fig. 3-1). At even higher treatment temperatures, the uptake of neutral red decreases to zero. The exact endpoint is hard to determine, in part because of the subjectivity of the test and in part because of the variability among plant cells, a confounding aspect of many biological measurements. Similar tolerance data occur for *Agave deserti* and *Opuntia ficus-indica*; the high temperatures decreasing the cellular uptake of neutral red by 50% for these two species growing at 30°C/20°C are 57°C (135°F) and 58°C (136°F), respectively (Nobel, 1988).

The high-temperature–tolerance curve has a similar shape when the plants are maintained for three weeks under hot day/night air temperatures of 50°C/40°C (122°F/104°F) instead of 30°C/20°C. However, the curve is shifted toward higher temperatures (Fig. 3-1). In particular, cellular uptake of neutral red by *F. acanthodes* is then not decreased by 50% until 67°C (151°F), a considerably higher such temperature than the 58°C occurring for plants maintained at 30°C/20°C. This ability to tolerate higher temperatures when the plants are maintained under higher temperatures is termed *hardening* or *acclimation*. It means that the plants have made cellular adjustments to protect against the extreme temperatures.

The acclimation is relatively rapid, as two-thirds of the overall adjustment to high temperatures by cells of *F. acanthodes* occurs in just three days (Fig. 3-1). Thus, plants can acclimate in a matter of days as the weather changes. Such acclimation is a key feature in dealing with temperature extremes. Moreover, all fourteen species of agaves and all eighteen species of cacti tested in this manner show high-temperature acclimation. The acclimation for these thirty-two species averages 9°C (16°F) when the day/night temperatures are raised from 30°C/20°C to 50°C/40°C (Nobel, 1988). For instance, the acclimations as the ambient temperatures are raised by 20°C are 7°C (13°F) for *A. deserti* and 9°C (16°F) for *O. ficus-indica*. Also for these two species, about two-thirds of the acclimation occurs in just three days, as for *F. acanthodes* (Fig. 3-1).

The hemiepiphytic cactus widely cultivated for its fruits, *Hylocereus undatus* (Chapter 1), does not tolerate high temperatures as well as its desert cousins. For instance, neutral red uptake by its chlorenchyma cells for plants under moderate growth temperatures but restricted water levels is reduced 50% at 55°C (131°F). Actually, severe stem damage of this species can occur below 50°C (122°F) when the plants are well watered and heavily fertilized (Nerd, Tel-Zur, and Mizrahi, 2002).

More importantly with regard to the highest temperatures that *H. undatus* can tolerate, its high-temperature acclimation as day/

night temperatures are raised by 20°C is less than 3°C (5°F; Nobel and De la Barrera, 2002a). This is much lower than the average high-temperature acclimation of other agaves and cacti (about 9°C). Nevertheless, its high-temperature tolerance is still greater than that of most vascular plants. Also, its cultivation under cloudy, artificially shaded, or naturally shaded understory conditions will not cause it to be exposed to particularly high temperatures, even taking into account the temperature increases expected during the 21st century in most regions (Chapter 4).

In summary, two features deserve special emphasis with respect to high temperatures: (1) many agaves and cacti can tolerate extremely high temperatures, and (2) most species of both taxa demonstrate a great ability for high-temperature acclimation. Thus increasing global temperatures are not really a survival threat for them, another crucial aspect in their potential for increased use in the future. The effects of increasing temperatures on the net CO_2 uptake and the productivity of agaves and cacti are discussed in Chapter 5.

More Incredible Survival Records

The soil surface can reach the threateningly high temperature of 70°C (158°F) in deserts, where perennial plants obviously survive. This is de facto evidence that any of their living tissues in contact with such soil can tolerate temperatures near 70°C. This is also true in principle for species of *Lithops* mentioned above, although the highest temperatures that they are exposed to in Namibia and in South Africa are not quite so high.

With regard to high temperatures, certain "dwarf" cacti from the Chihuahuan Desert of northern Mexico and southern Texas, such as *Ariocarpus fissuratus* (commonly known as the "living rock cactus"), *Epithelantha bokei*, and *Mammillaria lasiacantha*, face a similar situation as *Lithops*. Hence their high-temperature tolerances were investigated using the neutral-red technique. These species, which can be about 3 centimeters (1.2 inches) in diameter, can survive three days of day/night air temperatures of 58°C/48°C (136°F/118°F), which is extremely hot. Moreover, the stems of *A. fissuratus* can survive a one-hour high-temperature treatment of 71°(160°F; Nobel et al., 1986).

Persons go to great extremes to get their names in the *Guinness World Records*, such as having the longest hair (5.6 meters, 18 feet) or the longest kiss (nearly 31 hours, although some dispute that record). Speaking of records, so far we can crown *A. fissuratus* as able to tolerate the highest temperatures among agaves and cacti. To try to unseat this champion, various cacti were placed in environmental chambers with the hot day/night temperatures of 50°C/40°C; field observations were also made. Surprisingly, the ubiquitously cultivated *Opuntia ficus-indica* (Chapter 1) had chlorenchyma cells that could also tolerate 71°C when droughted plants were slowly acclimated to high air temperatures. The race is still on to decide the high-temperature champion.

The maximum temperature that can be tolerated for one hour by agaves or cacti is predicted to be 74°C (165°F) when properly acclimated and appropriately stressed. In particular, sometimes proteins that are induced by one stress can help cope with another stress, which is the case for the so-called "heat shock proteins" produced in response to high temperatures. In this regard, plants exposed to drought tend to be more tolerant of high temperatures than are well watered ones, as we have already indicated. Without going into the molecular or the genetic details, we simply note that this is an active research area. Again, our bottom line is that many agaves and cacti can survive high temperatures that are simply lethal to nearly all other plants, especially those used commercially.

Low Temperatures—Freezing a Water Bag

Although there is great news about the tolerances of agaves and cacti to high temperature, there is bad news regarding their tolerances to low temperatures. A bag of water will freeze when temperatures fall below 0°C (32°F). Thus freezing temperatures are generally an anathema to agaves and cacti. We mentioned (in Chapter 1) that *Hylocereus undatus* is unhappy even with chilling temperatures of 5°C (41°F). On the other hand, a few cacti are native to British Columbia, Manitoba, and Saskatchewan, Canada, where temperatures annually can fall below –40°C (–40°F; this is the only temperature that has the same numerical value on both temperature scales). For instance, the small prickly pear cactus *Opuntia fragilis* occurs at up to 57° north latitude in Canada and can survive such freezing temperatures, mainly by a remarkable acclimation ability for low temperatures (Loik and Nobel, 1993a).

Thus the genetics of cacti do not rule out tolerances of freezing temperatures, an aspect that we will return to at the end of this chapter. Also, the agave *Agave utahensis* is native to regions of northern Arizona, southern Nevada, and southern Utah where wintertime temperatures can be −25°C (−13°F; Nobel, 1988). Tolerance of freezing temperatures by both agaves and cacti is enhanced by drought, a common stress that biochemically predisposes the plants to tolerate other stresses.

Cooling curves and Cell Viability

When the air temperature surrounding a cactus stem is lowered at a rate that can occur in the field, the stem also decreases in temperature (Fig. 3-2). Because of its thermal inertia, however, the stem lags behind and remains above the air temperature during the cooling process. For the same sort of reason, the chlorenchyma cells on the outside of the stem drop below freezing before the internal water-storage cells do as the air temperature is lowered. This initial exposure to freezing cellular temperatures does not cause freezing damage, as the chlorenchyma cells contain solutes, which lower the temperature where freezing occurs. Specifically, as the solute concentration and hence the osmotic pressure of a solution increases, the temperature at which freezing occurs decreases by 1.86°C (3.35°F) per unit of molality (a one molal solution contains one mole of osmotically active solutes per kilogram of water).

The chlorenchyma cells of a well hydrated cactus stem may contain a 0.3 molal internal.solution, for which freezing would occur at (0.3)(1.86) or 0.6°C (1.0°F) below freezing. A very drought-stressed and hence dehydrated cactus may have chlorenchyma cells containing a 1.2 molal internal solution, which would freeze at (1.2)(1.86) or 2.2°C (4.0°F) below freezing. This is one aspect why drought-stressed agaves and cacti can tolerate lower temperatures than hydrated ones. Two other aspects of freezing tolerance relate (1) to proteins and protective solutes that are induced by the low temperatures and (2) to a shift of water out of certain cells, which we will consider shortly.

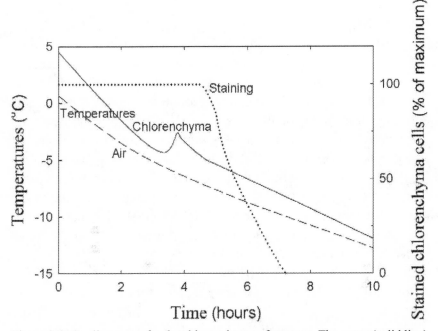

Figure 3-2. Cooling curve for the chlorenchyma of a cactus. The curve (solid line) indicates the decrease in temperature of the chlorenchyma as the air temperature (dashed line) decreases, its sudden increase in temperature after the supercooling event, which indicates that the freezing of water is taking place, and the subsequent resumption of a decrease in chlorenchyma temperature as the air temperature is lowered further. Also shown is the concomitant uptake of neutral red into the chlorenchyma cells (dotted line), which occurs for living cells only (see Fig. 3-1 for details of measurement). The rate of decrease of air temperature is similar to that occurring in the field. Staining is halved at an air temperature of about $-7°C$ (19°F), suggesting that plant death would occur at about $-10°C$ (14°F). The data are representative of those for various cacti and agaves (Nobel, 1988).

As the air temperature is decreased further below freezing, the temperature of the chlorenchyma suddenly increases (Fig. 3-2). What in the world is heating the cells? We all know that it takes energy (heat) to melt ice. Analogously, heat is released when water freezes. So this temperature rise indicates that water is freezing. Prior to this evidence of freezing, the cactus stem had been cooled below the temperature where freezing can occur. This is a non-equilibrium (metastable) process called *supercooling*. In particular, the internal solution supercooled below the temperature where it would normally freeze based on its solute concentration, but it did not freeze. Freezing began later, releasing heat and causing the chlorenchyma temperature to rise (Fig. 3-2).

We also note that essentially no chlorenchyma cells have died so far as the temperature is lowered (Fig. 3-2). Specifically, no decrease occurs in the uptake of neutral red, as the cellular temperature is first taken below freezing, during the subsequent supercooling event, and the initial stages of freezing itself. It is not until the stem has been cooled even further that we see any evidence of the decrease in the uptake of neutral red, indicating cell death (Fig. 3-2). A 50% decrease in the uptake of neutral red can occur for a chlorenchyma temperature of $-6.0°C$ ($21.2°F$), and stain uptake can be abolished at a chlorenchyma temperature of $-8.6°C$ ($16.5°F$). No stain uptake indicates cell death, and all cells being dead means that the plant is dead. This technique of examining the cellular uptake of a vital stain uses only small pieces of an agave leaf or a cactus stem. Thus it allows us to predict what freezing temperature will kill the plants without sacrificing any of them to get this information, as was similarly indicated when discussing high-temperature tolerances.

We often attribute the death of agaves and cacti at low temperature to the freezing of water, but actually their death is usually caused by cellular dehydration. The freezing that occurs leading to the chlorenchyma temperature rise (Fig. 3-2) is for tiny ice crystals located in the tissue but outside of the cells. Namely, these crystals are in the intercellular air spaces between the chlorophyll-containing chlorenchyma cells. In time, these tiny ice crystals continue to grow and grow as water diffuses out of the cells and becomes part of the enlarging crystals. Occasionally these growing crystals puncture the chlorenchyma cells, thus killing them. More usually, the cellular dehydration accompanying the loss of water from inside the cells to the extracellular ice crystals damages intracellular proteins and membranes, thereby killing the cells.

Even though many species of agaves cacti can withstand considerable dehydration, most species do not have the suite of characteristics necessary to tolerate really sub-freezing temperatures. *Opuntia fragilis* and *Agave utahensis* are exceptional in this regard. Thus the outstanding ability to tolerate drought generally is not accompanied by adaptations within the chlorenchyma cells of agaves and cacti to tolerate really low temperatures. Special adaptations to low temperatures, which can occur for *O. fragilis* (Loik and Nobel, 1993b), include the synthesis of specific proteins and protective solutes as well as hormones, which can counteract the effects of the major shift of cellular water to the intercellular air spaces during freezing.

Acclimation and Tolerances

Now that we have some idea of how agaves and cacti respond at the cellular level to freezing temperatures, let us consider their low-temperature acclimation ability. As for high-temperature tolerance, the best tolerators of low temperatures generally are the species having the best low-temperature acclimation abilities. That is, those species showing the greatest decrease in the temperature causing a 50% reduction in the uptake of neutral red as the day/night are temperatures are lowered usually are the best tolerators of freezing temperatures.

The current champion for low-temperature acclimation among cacti is *Opuntia fragilis*. It is certainly not fragile for low temperatures, as *O. fragilis* can harden or acclimate by an amazing 17°C (31°F) as the day/night air temperatures are lowered by 20°C (36°F; Loik and Nobel, 1993a). Among agaves, some of the best acclimators are *Agave parryi* and *Agave utahensis*, which can acclimate by 9°C (16°F) as the temperature is similarly lowered. For the thirty-five other species of agaves and cacti that have been tested in this manner, such acclimation averages only 1.1°C (2.0°F) for a 20°C decrease in ambient air temperatures. The reasons for the differences among species in the acclimation ability is not understood, but the poor low-temperature acclimation ability of most agaves and cacti underlies their susceptibility to freezing temperatures. Next we consider some morphological factors that can ameliorate the effects of low temperature.

We have mentioned that the temperatures of small plants at ground level are very similar to the temperatures of the soil surface there. The opposite is true of massive cacti, especially away from the ground. Moreover, shading by spines and a layer of hair-like cellular projections (called *pubescence*, as it resembles hair) at the top of a barrel cactus or the top of a columnar cactus can influence the stem temperature in this region (a cactus having pubescence is just like you wearing a furry cap to keep your head warm on a cold winter's night). The region at the top of a stem is biologically crucial for growth and survival, as it contains the *apical meristem*. The apical meristem is where cell division is occurring, which leads to the extension (elongation) growth of the stem.

Computer models were developed to predict the temperatures of the apical meristem for the barrel cactus *Ferocactus acanthodes* and the columnar cactus *Carnegia gigantea* in the Sonoran Desert of Arizona and California (Nobel, 1988, 2009). The calculations confirmed the importance of spines and pubescence in ameliorating temperatures at the top of the stem. For instance, such shading can raise the low tissue temperatures in the winter by 5 to 7°C (9 to 13°F), greatly extending the range of these plants to higher elevations and to higher latitudes compared with a stem having a bare top facing a cold nighttime sky. I can personally attest to this. Needing a slight trim, I went into a barbershop near Mexico City and, pointing to my head, said "Solamente en poquito" (meaning to me, "cut only a little bit"), The barber misunderstood and proceeded to shave my entire head. Boy, was the clear sky cold that night!

To test such computer models developed in North America, measurements were made on the barrel cactus *Eriosyce ceratistes* (*Eriosyce aurata*) and the barrel cactus *Trichocereus chiloensis* (*Echinopsis chiloensis*) in South America, namely in Chile. Field work involved measuring coverage of the stem by spines, apical pubescence, and stem diameters as well as searching for the highest elevation, and hence coldest habitats, for these species in the Andes Mountains.

Using binoculars, I spotted a suitable population on a distant hill, but it was across a narrow gorge that had a frothing river about 20 meters (65 feet) below. And there were two rusty cables about 1.5 meters (5 feet) apart stretching across the gorge. I said "Let's try them." The two scientists accompanying me took a few steps first, with their boot heels hooked on the lower cable and the upper cable under their armpits, but said "No way." If you fell, it meant going into ice-cold snowmelt in a raging river full of rocks. Although scared, I went across, made the measurements, and returned. Out of fear, I had gripped the upper cable so hard that it cut through my heavy jacket and two underlying shirts to the bare skin of my armpits. We then drove up the road to see if the highest population had been measured. Yes. But a short distance up the road was a real bridge across the gorge that made the foolhardy crossing totally unnecessary. Fieldwork is unpredictable!

All of the morphological parameters were fed into the computer model to predict the upper elevational limit of *E. ceratistes* and *T. chiloensis* in Chile. Amazingly, the upper elevations were predicted within 40 meters (130 feet) of the observed values for similarly exposed sites over a north-south distance of nearly 700 kilometers (over 400 miles). This represents a sensitivity of about 0.2°C (0.3°F) for the cactus populations with respect to the temperatures of the apical meristem. This also supports the contention that the apical meristem is the most sensitive part of these cacti with respect to survival (Nobel, 1980, 1988).

Getting back to the main story, the tolerances to low temperatures of most agaves and cacti of commercial importance (Chapter 1) is limited compared with various other perennials. Even given a chance for gradual acclimation, a temperature of −10°C (14°F) is lethal to most of them. Certain ornamental agave and cactus species are especially sensitive to freezing temperatures. Hence they are brought indoors during the winter for many personal gardens and botanical gardens of northern Europe, northern United States, Canada, northern Japan, much of the rest of Asia, and other such regions.

Let us return to the CAM plant with the greatest current region of cultivation and the most commercial importance, *Opuntia ficus-indica*. There are a large number of varieties and cultivars of this species—more than 100 worldwide. These have slight genetic differences and hence somewhat different environmental responses, representing a huge genetic resource. But a widely used cultivar has a low-temperature tolerance of −10°C (14°F). This means that only approximately 36% of the land area of California, the leading agricultural state in the United States, is currently suitable for its cultivation (Nobel et al., 2002); most of the exclusion occurs in mountainous regions with very low wintertime temperatures. Unfortunately, most of the land area appropriate for cultivation of *O. ficus-indica* is already claimed by high-value uses in and near cities and towns, especially along the Pacific Ocean coast, and for crops with higher commercial value.

Primarily because of its poorer tolerance of freezing temperatures (–2.5°C, 27.5°F), the area suitable for possible cultivation of *Hylocereus undatus* is only 2% of California's land area (Nobel et al., 2002). Most of this suitable land is in the southern coastal region, again in competition with high-value land-use demands. Thus the prospect of substantial increased area for the cultivation of this species under current climatic conditions in California is marginal, although the profits can be quite high per unit land area.

The increased usage of these cacti in California in the future requires consideration of climate change (Chapters 4 and 7) and physiological responses thereto (Chapters 5 and 6). For instance, changes are already occurring in the California agricultural industry that are presumed to be related to global climate change, such as a decrease of the time when chilling temperatures (below about 7°C or 45°F) occur in various regions. Thus the cultivation of apples, cherries, pears, and other crops requiring chilling temperatures is decreasing in the Central Valley where most of California's tree crops are grown. This can open up possibilities for other crops, such as agaves and cacti. Let us next consider another change possible in the future concerning the generally poor freezing tolerance of agaves and cacti.

Better Breeding and Better Biotechnology
We used the many different types of *Opuntia ficus-indica* to hint at genetic diversity at the species level. Similarly, the ability of *Opuntia fragilis* to tolerate –40°C hints at the genetic diversity at the genus level. What if we could exploit these traits through breeding programs or modern biotechnology? Certainly we could then attack one of the major shortcomings for cultivating agaves and cacti on a wider scale, the limited low-temperature tolerance of the commercially more important species and varieties.

Breeding, which is highly advanced for annual crops such as corn (maize), rice, soybean, and wheat, has had only limited application to agaves and cacti. This is due to various factors, including: (1) the longer time for these perennials to reach the reproductive stage, (2) their lower economic importance, and (3) their more limited geographical distribution. Thus many of the agaves and cacti used today have simply been "domesticated." That is, nicer specimens have been collected from the wild, propagated, and then utilized with little attention to pollinators, environmental responses, or molecular biological aspects. Management practices have indeed been perfected for *O. ficus-indica* and *Hylocereus undatus* for fruits and *Agave tequilana* for tequila, but this is really an exception to the rule. Further agronomic progress can even be made with these three species.

Breeding and biotechnology both seek to develop crops better suited to human needs. Improvements, such as developing more cold-tolerant cultivars, is best served if the genetic basis for this tolerance is understood. Other improvements for opuntias can relate to increasing fruit size, reducing the presence of seeds, and eliminating the glochids on the fruits (Chapman et al., 2002). Also, breeding and biotechnology can improve the nutritional value of cladodes used for forage or fodder.

Luther Burbank got credit for breeding a spineless *O. ficus-indica* through hybridization techniques (Chapter 1). More recent breeding programs have been directed toward fruit color and quality in Mexico (many governmental and private institutions involved), United States, Italy, Israel, South Africa, Argentina, Brazil, and Chile. Many of these studies examine the proteins and the sugars underlying the improvements. Also, some progress has been made in obtaining a genetic map for chromosomes of agaves and cacti, which can help identify molecular markers correlated with freezing tolerance.

In the future, a gene conferring freezing tolerance could be integrated into the germplasm of agaves and cacti. Indeed, breeding and biotechnology offer great promise for overcoming the low-temperature susceptibility and other problems holding back the increased commercialization of these CAM plants. Moreover, genetic analysis and molecular research on these champions of drought and high temperature can benefit other species as well. Who knows, perhaps agaves and cacti have held secrets for millions of years that may be crucial for the future of C_3 and C_4 crop plants as the climate changes.

4

Issues of Global Climate Change

The Atmospheric CO$_2$ Story

Most people recognize that atmospheric CO$_2$ levels are increasing. The data from Mauna Loa, the large volcano on the big island of Hawaii, accurately portray such CO$_2$ levels (Fig. 4-1). The increase from 1957 to 1999 averaged 1.4 parts per million (ppm) per year (*ppm* generally refers to a volume basis, such as the volume of CO$_2$ per total volume of air, which is often represented by *ppmv*; essentially the same ppm values occur for molecules of CO$_2$ per total molecules of air). More recently, the annual increases have become greater. For example, from 2000 to 2010 they averaged 2.1 ppm per year (Fig. 4-1). Thus the rate of increase in atmospheric CO$_2$ level is increasing. The repetitive annual oscillations in the Mauna Loa records also show the remarkable seasonal trends.

Having always been fascinated by the drama of volcanoes and especially contemporary lava flow, I remember the relatively short but ear-popping drive from about sea level in Hilo to visit Mauna Loa (peak at 4,170 meters, 13,680 feet) in 1970, when the records that would revolutionize our knowledge about atmospheric CO$_2$ levels had been accumulating for about a dozen years (Fig. 4-1).

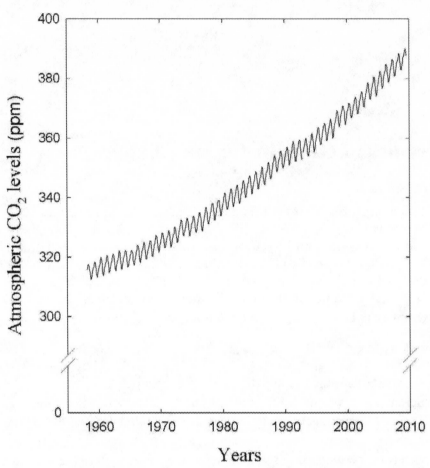

Figure 4-1. Measurements of atmospheric CO_2 levels at Mauna Loa, Hawaii. The NOAA site, which has been operating since 1957, is essentially free of local human perturbations with respect to CO_2. Therefore, it accurately portrays the recent changes in global CO_2 levels, including the spectacular seasonal variation of CO_2 levels in the Northern Hemisphere. Data are based on dry air, i.e., after the water vapor has been removed. [Adapted from www.mlo.noaa.gov.]

Mauna Loa is the world's largest shield volcano (meaning shaped like a shield), whose last major eruption was in 1984. In 1957 the United States National Oceanic and Atmospheric Administration (NOAA) established the Mauna Loa Observatory at an elevation of 3,400 meters (11,100 feet) to monitor atmospheric CO_2 levels. This site is well isolated from human influences coming from Hilo and from local industrial activities (outgassing of CO_2 from volcanic activity is also corrected for). But let us get back to the CO_2 observations.

In this regard, the atmospheric CO_2 level in the Northern Hemisphere decreases during the summer, when plants there are growing well and hence are taking up CO_2. It increases during the winter, when there is a net release of CO_2 as respiration (plus photorespiration; Chapter 2) that exceeds photosynthesis. By themselves, details of these annual oscillations (Fig. 4-1) indicate the great sensitivity of the measurements. Moreover, because the magnitude of the annual oscillations exceeds the net annual increases, plants have a greater seasonal influence on atmospheric CO_2 levels than do anthropogenic emissions of CO_2 (more about this soon). Nevertheless, the annual changes in atmospheric CO_2 levels will have major and predictable effects on net CO_2 uptake by CAM plants, as detailed in the next chapter. Here we ask: "Are these changes real?," "Will they continue?," and "What is the historical perspective?"

The answers to these three questions are "Yes," "Yes," and "Let's see." Actually, the atmospheric CO_2 levels have varied widely over the years, as detailed from ice cores taken from Greenland and from Antarctica containing gases trapped eons ago. Although paleoclimates are interesting, such as what happened 700,000 years ago when atmospheric CO_2 levels were considerably lower than they are today (below 300 ppm), our main concern is more immediate. Over the last 12,000 years and up to the Industrial Revolution in about 1750, the atmospheric CO_2 level was relatively stable at 260 to 280 ppm. The approximately exponential increase in the atmospheric CO_2 level since then has led to the anointing of a new geologic epoch often referred as the "Anthropocene," emphasizing the current central role of humans in geology and ecology.

Specifically, at the beginning of the 18th century just before the Industrial Revolution, the atmospheric CO_2 level was about 270 ppm. It increased to approximately 280 ppm at the beginning of the 19th century. By 2010, the level had reached about 390 ppm (Fig. 4-1). And for the end of the 21st century, estimates average about 750 ppm. Various scientists and international political organizations have suggested capping the atmospheric CO_2 level at 500 ppm, which would involve drastic changes in how we do business today. For the purposes of this book, we will consider what such levels mean for plants. What they mean for politicians in particular and the public in general are much broader topics that are more fully explored elsewhere.

Before looking at the causes of the currently increasing atmospheric CO_2 levels (Fig. 4-1), let us comment on its relative stability for the 12,000 years preceding the Industrial Revolution. Stability means that the processes adding CO_2 to the atmosphere compensate the processes that are removing CO_2. As discussed in Chapter 2, photosynthesis removes CO_2, while respiration and photorespiration are two processes adding CO_2. Another process adding CO_2 is fire, both natural (such as caused by lightning) and anthropogenic. The chemical reaction for respiration and the burning of wood and other plant material is just the reverse of Equation 2.1, namely, *carbohydrate* plus *oxygen* reacts to form CO_2 plus *water*. Open fires from all causes currently release about 4% as much CO_2 into the atmosphere as photosynthesis takes up. Of course, megafires change the proportions locally, such as the forest fires burning more than 200,000 acres each (over 300 square miles or 800 square kilometers) occurring recently in the western United States.

Various sources are responsible for the currently increasing levels of atmospheric CO_2 and other greenhouse gases (greenhouse gases are defined in the next section when the global temperature is considered). The primary culprit for the increasing CO_2 is the burning of fossil fuels. (As an aside, what would we do in our day-to-day habits without coal, natural gas, diesel fuel, and gasoline?) Burning of fossil fuels releases approximately 6×10^{12} kilograms of carbon (in the form of CO_2) into the atmosphere annually, which is about 6% of the annual photosynthetic productivity of about 110×10^{12} kilograms of carbon fixed (presented at the beginning of Chapter 2; Chapin et al., 2002; Nobel, 2009). About half of the CO_2 released by the burning of fossil fuels leads to increases of the atmospheric CO_2 level, while approximately equal portions of the remaining half are absorbed by terrestrial vegetation and by lakes or oceans.

Realizing that our own respiration releases CO_2 into the atmosphere, an environmentally conscious student once asked me: "How much does our collective breathing contribute to the increasing level of CO_2 in the atmosphere and the resulting global warming." Actually, the approximately 7 billion humans on Earth in 2010 contribute only about 0.7×10^{12} kilograms of carbon annually via respiration (less than 1% of the CO_2 fixed by photosynthesis). So holding our breaths will have little effect on global warming.

In terms of anthropogenic CO_2 emission, people also point at cement manufacturing. It releases about 0.9 kilogram of CO_2 per kilogram of cement made (both from the burning of fossil fuels during processing and the chemical process itself, which causes the carbonate present in rocks used to be released as CO_2). This is a relatively small emission source, accounting for less than 2% of that from other uses of fossil fuels. A much more important source is land-use changes, which currently release nearly one-third as much CO_2 into the atmosphere as the burning of fossil fuels. Land-use changes often involve replacing forests, with much CO_2 stored in the biomass, by agricultural crops, with little permanent standing biomass. This is currently occurring in Brazil, Indonesia, parts of Africa, and other locations worldwide. Livestock husbandry often can mean replacing forests having much standing biomass with grasslands having relatively little biomass, with similar consequences for decreasing terrestrial CO_2 storage.

Changing the worldwide emissions of CO_2 consequently involve land-use decisions as well as switching from the dependence on fossil fuels to other power sources that do not generate CO_2, important and interesting topics. Thus, much political and public debate centers on the economic consequences of responding to the observable increases in atmospheric CO_2 levels. This book focusses on the biological consequences of the clearly documented increasing global atmospheric CO_2 levels—in particular, for biomass productivity and specifically for CAM plants such as agaves and cacti.

A moderate setback in the study of global atmospheric CO_2 levels occurred on February 24, 2009, when a satellite launching was unsuccessful. Shortly after takeoff from Vandenberg Air Force Base in California, a heat-shield failed to separate from the $278 million (U.S. dollars) satellite launched by the National Aeronautics and Space Administration (NASA), which was established in 1958 to oversee the space program in the United States. The satellite carried spectrophotometers designed to measure CO_2 emissions and to identify where natural processes were sequestering carbon, such as in the oceans and in forests. Indeed, more knowledge is needed concerning such carbon "sinks" (discussed in Chapter 7) to effectively combat the increasing atmospheric CO_2 levels. Also, the "Orbiting Carbon Observatory" carried by the satellite was to have seen what effects current efforts in reducing CO_2 emissions were having, which could have had major political and economic consequences.

Let us also examine a few more facts about CO_2. The CO_2 in the atmospheric is recycled every 3 or 4 years by its removal via photosynthesis and its return via respiration and photorespiration. That is, the metabolism of animals, plants, algae, bacteria, and fungi controls the short-term fluctuations in atmospheric CO_2 levels (Fig. 4-1). Superimposed on this are the longer-term anthropogenic CO_2 emissions. These processes lead to exchanges between the atmosphere and the oceans, where approximately 50 times more carbon in the forms of CO_2, bicarbonate, and carbonic acid reside than carbon as CO_2 in the atmosphere.

Increasing temperatures drive the gas CO_2 out of the oceans, contrary to the case for dissolved solutes, whose solubility increases with increasing temperature. CO_2 is also used to produce the bubbles in carbonated soft drinks. And natural fermentation leads to the CO_2 bubbles in beer and in champagne, without which this would be a much more dreary Earth! Again, warming such beverages leads to an observable release of gaseous CO_2.

Temperature—It's Getting Warmer

In terms of broad brush strokes, environmental temperatures are increasing. Such increases have been harder to prove than atmospheric CO_2 increases (Fig. 4-1), because (1) the magnitudes of the changes are smaller, (2) precise comparative temperature measurements are more difficult, and (3) techniques have changed over the years.

For instance, the oceans represent huge bodies of water that are expected to have relatively stable temperatures. So why did some measurements for ocean temperatures suddenly jump about 0.5°C (0.9°) during the 19th century? Well, one temperature-measuring procedure used in the early 19th century was to drop a bucket over the side of a sailing ship, hoist it up, and stick in a thermometer. As steam ships came into being in the latter half of the 19th century, water was more easily taken from the propeller shaft, which heated it slightly, leading to an artificial temperature increase estimated for the upper Atlantic Ocean. Also, many weather stations on land are situated near cities or airports, where increased urbanization, exemplified by new buildings, new asphalt roads, parking lots, and high local energy use, biases long-term records (the "urban heat-island effect").

We can still consult the most reliable measurements and correct for known artifacts. We then find that air temperatures have tended to increase beginning in 1900 compared with the previous 900 years. During the latter quarter of the 20th century and on into the 21st century, global air temperatures have been rising at about 0.19°C (0.34°F) per decade. Although this may seem small, it is a unidirectional trend that so far is unstoppable. Moreover, this trend is expected to continue and actually to increase for a long time. We can also look at this trend in more down-to-earth terms by saying that spring is coming about three days earlier each decade (Smith, Saatchi, and Randerson, 2004).

The current annual increases in atmospheric CO_2 and other greenhouse gases, such as methane, are known. These gases lead to an increased trapping of heat in the lower atmosphere, known as a "greenhouse" effect and leading to the name *greenhouse gas*, which affects global temperatures. Everyone knows what greenhouses are, although many in the United Kingdom more appropriately call them glasshouses, as they are generally covered by glass and are not green. Nonetheless, these structures for growing plants "indoors" have temperature effects somewhat analogous to but actually not truly the same as our Earth's atmosphere. In particular, greenhouses prevent air escape, or convective heat exchange, whereas the atmosphere involves mainly radiation heat exchange based on its molecular constituents, so the "greenhouse" terminology is not strictly correct.

Greenhouse gases such as CO_2, methane, water vapor, and ozone absorb and emit longwave (infrared) radiation (also known as thermal radiation), as does the surface of the Earth. The radiation emitted by greenhouse gases goes in all directions, so their increasing concentration in the lower atmosphere means more radiation is directed back toward the Earth (like the restrictions of a greenhouse roof, but for different reasons, as already mentioned). This causes the global air temperature to increase. Actually, the radiation of longwave radiation back to the Earth by atmospheric gas molecules is crucial for maintaining life-sustaining temperatures on the Earth, so it is more a quantitative effect rather than a qualitative one.

Now that we have defined greenhouse gases, a few words about methane. This small molecule, with the chemical formula CH_4, is about 22 times more potent per molecule than CO_2 in trapping heat in the lower atmosphere. Methane is emitted in rice cultivation, marshes, certain mining operations, most landfill operations, termites, and notoriously by the flatulence and belching of cattle. Cattle and other ruminants release approximately 110×10^9 kilograms (110,000,000 tonnes) of methane gas annually. Imagine the fireball that would result if all of this methane gas were placed in a single cloud and then somebody lit a match nearby! But such burning of methane would actually lead to less long-term global warming.

Specifically, combustion of a molecule of methane produces a molecule of CO_2, which leads to only (100%)/(22) or 5% as much global warming as methane. Natural plus anthropogenic emissions of methane total about 600×10^9 kilograms annually. Methane levels in the atmosphere tripled from 1850 to 2010, when they approached 2 ppm. An unpredictable and potentially more threatening future event is the release of methane (1) from the melting of permafrost, (2) from undersea vents, and (3) from submerged crystallized methane hydrates. Because methane is much more potent than CO_2 as a greenhouse gas, incorporating these future methane emissions is particularly crucial for climate models. Indeed, the United Nations Environment Programme called new methane release "the global warning wild card."

Other atmospheric components that influence air temperature are anthropogenic aerosols and those due to volcanic eruptions, which actually have cooling effects. "Aerosol" refers to fine solid particles or tiny liquid droplets suspended in the air. Anthropogenic sources of aerosols include smoke, smog, and air pollution in general, as well as sulphate-containing droplets released from the burning of fossil fuels. The cooling relates to "global dimming" due to the reflection of sunlight away from the earth by the aerosols. Suggestions have even been made to reverse the global temperature rises by releasing more such aerosols into the atmosphere.

Aerosol effects are now included in climate models. However, volcanic eruptions create special difficulties for modelers because so far they and their aerosols are unpredictable in time. The initial cooling of the Earth's air temperature by volcanic eruptions is due to particulate matter injected into the atmosphere, which can reflect incoming radiation and shade the earth's surface. Later cooling relates more to atmospheric changes caused by the emitted gases, especially SO_2, which can form sulphate aerosols. To complicate matters further, volcanic eruptions also release water vapor and CO_2 into the atmosphere, which can have heating effects.

Models (now more than 25 worldwide) are used to predict future temperatures. These models are collectively referred to as General Circulation Models, or GCMs. The more sophisticated ones are also called AOGCMs, where "AO-" refers to "Atmosphere–Ocean." More local models are called RCMs (Regional Climate Models), which give more precise information for a specific region, such as for the Sahel of Africa or for southern California. GCMs are basically sets of differential equations describing the motion of a fluid, in this case air. Their computational complexity requires today's high-speed computers, such as for forecasting the upcoming weather (some related good news is that the accuracy of weather forecasts has greatly improved over the years).

Let us next consider pronouncements with respect to global temperature made by the Intergovernmental Panel on Climate Change (IPCC), which shared the 2007 Nobel Peace Prize with Al Gore, the former United States Vice President who was cited for his publicizing the impending global climate change. The IPCC was established in 1988 by the World Meteorological Organization and the United Nations Environment Programme. Its purpose is to evaluate published scientific literature with respect to global climate change and to issue periodic summary reports. The Fourth

Assessment Report issued by the IPCC in 2007 had contributors from over 130 countries. The report stated with a greater than 66% confidence that there would be an increase in drought in the 21st century, and stated with a greater than 90% confidence that there would be an increase in warm spells and heat waves. Also, the warmer spring temperatures would be occurring earlier (also stated with greater than 90% probability).

GCMs evaluated by the IPCC and based on the rates of increases of greenhouse gases and a range of other assumptions predict an average increase in temperatures worldwide of 3.1°C (5.6°F) by the end of the 21st century (Christensen et al., 2007; Randall et al., 2007). To help put this temperature increase into perspective, an average increase of 3.1°C will shift the frost-free region about 350 kilometers (220 miles) poleward in both the Northern and the Southern Hemispheres. This can have major agricultural and ecological ramifications. In any case, a considerable range occurs in the estimated temperature increases predicted for 2100, most being between 2.2°C and 4.7°C (4.0 to 8.5°F). Uncertainty exists concerning future greenhouse gas emissions, the effects of clouds, changes in glaciers, and the inherent complexity of computer models, among other things. Hence even the degree of uncertainty among the estimates is uncertain.

To make progress regarding temperature predictions as well as for other future environmental conditions, in this book we will use the procedure adopted by the IPCC. Namely, we will use the average of the various models. The hope is that the possible errors and incorrect assumptions of the hundreds of scientists involved will cancel each other out. Consequently, their collective and extensive expertise will get the picture about right.

Local temperatures are more important for plant growth and for biomass productivity than are global average temperatures (regional implications for the future cultivation of agaves and cacti are considered in Chapter 7). In this regard, temperatures over landmasses are expected to increase more than are the global average temperatures (Christensen et al., 2007). This reflects less water available for evaporative cooling and lower inertia for temperature changes over the continents compared with over the oceans, which occupy about 70% of the surface area of the Earth and are convectively mixed.

Specifically, temperatures at the end of the 21st century are expected to rise more than the global average in the upper and the mid-Atlantic region, such as northwestern Europe (1.2°C above the average), northeastern North America (0.8°C above the average), and especially above the Arctic Circle (over 3.0°C above the average in the winter). A large part of the landmass of South America is in the Northern Hemisphere, where temperature changes similar to the average global changes are predicted. Similar average predictions occur for Australia.

Africa may be a key region for increased future biomass productivity of agaves and cacti. But limited resources have handicapped local data acquisition and hence model testing (particularly acute for western Africa). Most of this continent lies in the Northern Hemisphere but near the Equator, where land and global predicted temperatures are similar. For Africa, temperatures in the Sahara in 2100 are predicted to be about 0.7°C above the global average and about 0.4°C above the global average for the rest of the continent. Location within a continent is also important, as temperatures are predicted to rise more in the interior of continents than on their periphery where contact with oceans occurs.

Clouds, Rainfall, and Models Again

Throughout this analysis of global climate change, clouds remain an immense problem, both for interpreting existing data and for models predicting future climates. Clouds are whimsical ... developing where water vapor condenses and disappearing where it evaporates, often within hours at the same site. Clouds reflect the sunlight impinging on their tops away from the Earth, which is a cooling effect. Clouds also emit heat in the form of longwave (infrared) radiation toward the Earth, which is a warming effect. Water vapor in clouds, and in the atmosphere in general, is the most important greenhouse gas with respect to global temperatures—to the consternation of climate modelers. Moreover, clouds not only dictate local temperatures but also affect local rainfall patterns, which are crucial for plants.

Although considerable agreement has been reached on the magnitude of the global CO_2 increase and the accompanying temperature increases, predictions of changes in rainfall patterns are less precise and yet in many ways are more critical for plants. For the southwestern United States, models predict a decrease of annual rainfall and a shift from major winter rainfall to a more stochastic (variable) mix during the year. Instead of relying solely on such models, let us examine the historical record for southern California over a period of 132 years. Although everybody says that the weather is getting worse, no striking changes are generally in evidence. Yet the average annual rainfall over the last 66 years in Los Angeles is 9% less than over the previous 66 years (the years considered embrace the total extent of the rainfall records for the Civic Center of Los Angeles from mid 1877 to mid 2009). Is this a trend? Probably too soon to tell.

Next we reconsider the models used to predict future temperatures and rainfall patterns. Such AOGCMs include topographical features such as mountains, water released into the atmosphere by plant transpiration, CO_2 released into the atmosphere by anthropogenic emissions, changes in the Southern Hemisphere ozone hole, influence of sea ice and land ice sheets, and anything else the groups of modelers think should be included. Thus modeling is both an art and a science, because the outcomes depend on assumptions about future human behavior as well as the laws of physics and chemistry.

Also unpredictable events, such as massive volcanic eruptions can occur. Such eruptions can lead to large-scale cooling of the Earth's atmosphere, as indicated in the last section. Future methane release is also unpredictable, as also mentioned in the last section. In addition, the radiation output of the sun varies, such as the 11-year cyclic changes accompanying sunspots. Increases in global temperature lead to increases in the spatial variability of rainfall. Decreases in precipitation are predicted for subtropical regions and increases are predicted for parts of the tropics and for higher latitudes in the Northern Hemisphere during the 21st century.

Modelers have to worry about (1) heat exchange based on water evaporation (called "latent" heat), (2) heat exchange based on conduction and convective air movement (called "sensible" heat), and (3) heat exchanges based on radiation (such as that emitted by solid surfaces as well as by molecules in the air). Take glaciers, which we will discuss shortly, as an example. Their whitish color, which means much reflection of sunlight (a high albedo), can be replaced by the brown color of the underlying rocks (a low albedo) when they retreat. Evaporation and sublimation of glacial water/ice injects water vapor, a potent greenhouse gas, into the atmosphere. Glaciers also affect the temperature of the oceanic water, with the ocean surface temperature being an important variable in many AOGCMs.

Choices are also necessary for the spatial resolution of predicted future climates. For instance, the atmosphere can be vertically divided into 20 layers, each considered individually. Horizontally, a model can be divided into regions of 1 or 2° of latitude (1° of latitude corresponds to 111 kilometers or 69 miles) and 2 or 3° of longitude (1° of longitude equals 1° of latitude at the equator but becomes vanishingly small at the North Pole or the South Pole). Some high-resolution climate models use a spatial resolution of 0.2° or even less, which enormously increases the number of calculations involved (even finer resolution is used in certain RCMs). Besides spatial resolution, the models must also consider temporal resolution, which for the outputs varies from hours to days for weather forecasts and from years to decades for future climates.

How do the models considered by the IPCC deal with rainfall patterns and amounts? At the micro level, some models consider raindrop formation. Such models are then used to analyze the distribution of liquid water and ice within clouds. The clouds are then allowed to move both horizontally and vertically. Other models deal with water vapor in terms of relative humidity. When the water vapor saturation content is reached (100% relative humidity; Chapter 2), condensation will occur, as in cloud formation. This saturation water vapor content increases approximately exponentially with temperature, showing the interrelationship among clouds, modeled air temperatures, and rainfall.

Coupling between the land and the atmosphere occurs when interactions between soil moisture and precipitation are included.

This is closer to home for our discussion, as water is fundamental to plants and their biomass production. Models are often tested against known weather patterns. In this regard, GCMs have difficulty simulating the increase in rainfall in the early evening observable in many locations. Many simulations yield more rainfall events but with less intensity than are observed. Errors tend to cancel each other so that the predicted seasonal precipitation is fairly realistic when compared with recent weather records.

Some effects of drought on agaves and cacti were considered in the last chapter, but let us anticipate a few future trends here (Christensen et al., 2007; Kolbert, 2009). Specifically, rainfall is predicted to increase near the equator, to decrease in subtropical regions near 30° north latitude and 30° south latitude, and to increase in more poleward regions. Thus, annual rainfall is expected to decrease in northern Africa in general and in the Sahara in particular. Decreases are also predicted for the adjacent Mediterranean region of Europe, whereas increases are predicted for northern Europe. For Asia, annual precipitation is expected to increase in most regions, but to decrease in India as well as in the Middle East. Precipitation is predicted to increase in Canada and northern plus northeastern United States but to decrease in the southern plus southwestern United States, parts of Mexico, and in central South America. Southern Africa and Australia should become drier, but New Zealand may become wetter.

Thus, the rainfall predictions vary a lot regionally. Also, better modeling is needed to improve the predictions. Taking both temperature and precipitation into consideration, the line in the Northern Hemisphere delineating where droughts might be expected based on precipitation minus water evaporation will be shifting poleward (northward) in the 21st century.

From Glaciers to the Gulf Stream

What are some of the other factors that should be considered with respect to global climate change? People love glaciers, and their retreats or advances are serious events locally. More than 8,000 glaciers are named and at least to some extent monitored annually worldwide. Glaciers occur on every continent, when those for Mount Kilimanjaro and a few other mountains are included for

Africa (nearly all of the glacial ice on Mount Kilimanjaro in 1912 has since disappeared).

Of the most intensively monitored glaciers, 95% have retracted since 1980. This includes nearly all of the glaciers in Europe, which have been studied in considerable detail. A case in the Western Hemisphere is Glacier National Park, located in northern Montana and established in 1910. Of the 150 glaciers in the Park region in 1850, only 27 remained in 2005. If the current global climate warming continues at its projected rate, essentially all of the glaciers in Glacier National Park will be gone by 2030, with only small patches of glacial ice remaining.

The retreat of glaciers over large regions can drastically affect the absorption and the reflection of solar irradiation, as indicated above. The data bank is enormous for glaciers and clearly relates to global warming. Glacial movements are also coupled to El Niño/ La Niña cycles, all kinds of other Earth cycles, and solar cycles. Their interpretation is challenging but perhaps the most dramatic and tangible evidence for global climate change.

When a glacier "calves" and thrusts an iceberg into the ocean, thereby moving ice from the land to the sea, the oceans gain more mass. This leads to a rise in oceanic water level. However, once floating an iceberg does not affect ocean levels as it melts, just as a melting ice cube in a drink does not affect the liquid level in the glass (ice is about 91% of the density of liquid water, so about 9% is above the water level; upon melting, the former ice volume shrinks and assumes the density of water, so the level does not change). Oceans are thus predicted to rise due to the movement of glacial land ice into the seas as a consequence of global warming. The effect of rising oceans on where agaves and cacti can be cultivated is minor, but the implications for human populations in coastal regions and river deltas is enormous.

What about the incredible thermohaline circulation in the oceans ("thermo-" refers to temperature and "haline" refers to the salt content or salinity)? This thermohaline circulation is driven by differences in seawater density. Will global climate change affect this most remarkable and even mysterious circulation pattern?

Imagine the improbability of a shallow water current beginning in the upper Pacific Ocean (Fig. 4-2). Warming as it moves west and south, the main portion passes north of Australia, warming more in the Indian Ocean, and then passes south of Africa before heading northward in the Atlantic Ocean. The main current then brushes northeastern South America before heading to the Caribbean and the Gulf of Mexico (Fig. 4-2). Next it moves basically northeast as the *Gulf Stream* and its extension, the *North Atlantic Drift*, heading to northwestern Europe. Later it cools and then subverts (goes under) as it has become saltier (hence heavier) due to water evaporation and the crystallization out of pure water (lower temperatures also increase its density). It now retraces its steps toward the Pacific Ocean but now most goes south of Australia (Fig. 4-2). Could the vulnerability of this unbelievable water and energy conveyor belt, which can bring mild temperatures to Europe, be a catastrophe in the making? Will melting of Greenland ice be the culprit in changing the water density and the flow pattern?

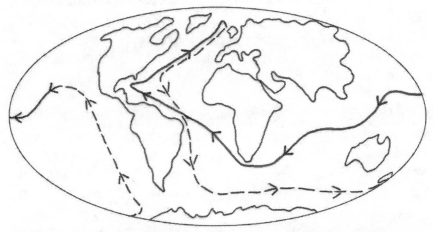

Figure 4-2. Overview of a major oceanic water current system. The deep water circulation just north of Antarctica and many other intersecting loops are not included in this simplified picture. The solid line indicates a surface current, and the dashed line indicates a deeply submerged current. The Gulf Stream and its extension, the North Atlantic Drift, can raise temperatures in northwestern Europe; this moderating effect on local temperatures is also highly dependent on global air currents. Such interactions of various convective and radiative heat exchange mechanisms are considered by AOGCMs. [Adapted from Rahmstorf, 2006; www. earthobservatory.nasa.gov.]

Although most scientists do not feel that the Gulf Stream in particular and the thermohaline circulation in general are immediately endangered, the stakes of a change are extremely high. Consider those headline-grabbing events that occurred in 2004—a United States Pentagon report revealed in *Fortune Magazine*, the 20th Century Fox movie "The Day after Tomorrow," and the British newspaper *The Guardian* report on a drastic local reduction in the North Atlantic Drift. All focused on an "abrupt climate change" that caused these oceanic circulation patterns to stop in a matter of days. Such abrupt changes more likely would occur over the course of a few years, which is a low-probability but still high-impact scenario. AOGCMs essentially all predict a diminished water circulation in the Gulf Stream and the North Atlantic Drift during the 21st century, although none predict their collapse. Such possibilities are incorporated into the models, but many consequences of global climate change are hard to predict.

As we stare into the crystal ball, what do we see? Oceans are rising, albeit at a slower rate than many have predicted. Atmospheric CO_2 levels and global temperatures are rising. Rainfall patterns apparently are changing, and hurricanes have increased in intensity from 1980 to 2010 compared with earlier times. The global temperature will most likely continue to rise after the 21st century due to the large amount of heat that can be stored in the oceans and the multiple-year turnover time for CO_2 in the atmosphere. We cannot predict exactly what will happen in the future, but no drastic threats from climate change are expected for agaves and cacti.

Plants Benefit from Increasing CO_2

What if you were to offer a banker more money? Or to offer an athlete more food? Or a plant more CO_2? All would say, "Yes, I'd like that!" In this regard, Equation 2.1 indicates that atmospheric CO_2 is fixed in plants by the process of photosynthesis. The higher the atmospheric CO_2 level, the higher is the rate of photosynthesis, at least up to a point. Plants have coped with various atmospheric CO_2 levels over geologic time. But the advent of increased release of CO_2 into the atmosphere since the beginning of the Industrial Revolution (i.e., during the Anthropocene) has been good news for them as far as photosynthesis and productivity are concerned.

Actually, the news varies with the photosynthetic pathway utilized, as we consider next.

Because the net rate of CO_2 uptake by leaves of C_4 plants is nearly saturated (maximal) at current atmospheric CO_2 levels, C_4 plants exhibit only small effects on photosynthesis when the atmospheric CO_2 level is experimentally raised from the current level of just below 400 ppm up to 700 ppm, an approximate doubling that is commonly used in research studies (Drennan and Nobel, 2000; Nobel, 2009). On the other hand, the rate of photosynthesis for C_3 plants generally increases at least 30% as the atmospheric CO_2 level is raised from 350 ppm to 700 ppm for short-term experiments lasting days to weeks, although the enhancement is generally less for experiments lasting months to years. The effect on CAM plants is variable, in part because they can take up CO_2 both during the daytime and during the nighttime (Fig. 2-4). In any case, the percentage enhancement in net CO_2 uptake at 700 ppm compared with 350 ppm for CAM plants during short-term experiments generally equals or exceeds the minimally 30% increase found for C_3 plants, and the enhancement tends to last for longer times (considered in the next chapter).

The current increases in atmospheric CO_2 levels are eroding the advantage of the CO_2 concentrating mechanisms in bundle-sheath cells of C_4 plants during the daytime and in chlorenchyma cells of CAM plants at night (discussed in Chapter 2). In particular, these plants have special means to raise the CO_2 concentration at the sites where Rubisco occurs (Fig. 2-1), thereby decreasing photorespiration. However, not until the atmospheric CO_2 level exceeds 1,000 ppm will the advantage of the biochemical concentrating mechanisms in C_4 and CAM plants be substantially compromised relative to C_3 plants, where CO_2 fixation is more straightforward in terms of the processes involved and their cellular location (again see Fig. 2-1).

Yet the plausible increase in photosynthetic rates by providing more CO_2 is only part of the good news for plants. If a higher concentration of CO_2 is present in the atmosphere, then the stomata do not have to open as much to let sufficient CO_2 diffuse into the leaves and the stems to support the ongoing rates of photosynthesis. If the stomata do not open as much, less water is lost by transpiration, which raises the Water-Use Efficiency (WUE, Equation 2-3). Thus increased atmospheric CO_2 levels are a win–win situation for the WUE of plants, as the amount of CO_2 fixed by photosynthesis increases and the amount of water lost by transpiration decreases.

Before we go into a few technical details, let us summarize what could happen to the Water-Use Efficiency of plants for an approximate doubling of the atmospheric CO_2 level to 700 ppm, ignoring other compensating or complicating responses. Specifically, the immediate WUE by C_3 plants would be raised by about 60%, that of C_4 plants by about 30%, and that of CAM plants by at least 60% (Nobel, 2009). The possible doubling of atmospheric CO_2 levels during the 21st century and beyond will be accompanied by increasing temperatures and by changing rainfall patterns, which also affect net CO_2 uptake, transpiration, and productivity in other ways. We will discuss these in the next chapter for agaves and cacti.

Because of the increased net CO_2 uptake, biomass productivity of CAM plants increases about 35% in response to an experimental doubling of the atmospheric CO_2 concentration (Drennan and Nobel, 2000; Chapter 5). The increasing Water-Use Efficiency leads to a higher proportion of the daily (24-hour) net CO_2 uptake to be Rubisco-mediated and to occur during the daytime. Also, daily net CO_2 uptake occurs for a longer period into drought under higher atmospheric CO_2 levels.

Root:shoot ratios for CAM plants tend to increase with atmospheric CO_2 level, reflecting the greater nutritional needs and hence more roots required accompanying the higher photosynthetic rates. Shoot morphology is also affected. For example, leaves of agaves and most other plants and the chlorenchyma for both agaves and cacti become thicker under elevated CO_2 levels, which serve to support the higher net CO_2 uptake rates. That is, more photosynthetic tissue with a greater internal surface area develops per unit external surface area of the leaves and the stems as the atmospheric CO_2 level increases.

Plant growth in general is enhanced by elevated atmospheric CO_2 levels. For instance, new cladodes develop sooner for *Opuntia ficus-indica* under doubled atmospheric CO_2 levels, and its biomass productivity then increases by 40%. Taken together, the anatomical, morphological, and physiological changes accompanying elevated atmospheric CO_2 levels are beneficial for plants in general and for agaves and cacti in particular. We must recognize, perhaps reluctantly, that there is some good news accompanying global climate change, at least as far as net CO_2 uptake, productivity, and cultivation of agaves and cacti are concerned.

Our CAM Plants Coping with Future Temperatures

If we were to design plants ready for globally increasing temperatures, we would pick those that can tolerate higher temperatures than those currently prevailing and that did not especially like the currently ambient low temperatures. Agaves and cacti!

As the previous chapter has indicated, agaves and cacti are remarkable among vascular plants in being able to tolerate extremely high temperatures. Their high-temperature tolerances and their high-temperature acclimation ability may contain secrets genetically useful for other crop species (mentioned at the end of Chapter 3). Again the good news does not stop here, as the susceptibility of CAM plants in general and agaves and cacti in particular to freezing temperatures can also be partially offset by higher air temperatures in the future. For example, we indicated earlier in this chapter that the average temperature increase predicted by the end of the 21st century (3.1°C, 5.6°F) can benefit agaves and cacti, many of which are averse to regions with frost, with 350 kilometers (220 miles) more poleward regions for cultivation (at a given elevation).

Although we will defer consideration of regional land-use specifics to Chapter 7, we need to re-emphasize some of the temperature survival aspects presented in the previous chapter. Agaves and cacti, when properly acclimated—meaning that air temperatures in experiments are gradually increased over a period of days or weeks—commonly survive 60°C (140°F). This is really special. Even the worst-case scenarios predict only a 5°C (9°F) increase in air temperatures by the year 2100. In that regard, how many regions currently experience temperatures above 55°C (131°F) that would eventually reach these high temperatures? Not many.

Moreover, some agaves and cacti can tolerate much higher temperatures than 60°C, notably, the widely cultivated *Opuntia ficus-indica.* Thus their future survival is not threatened by high temperatures. On the other hand, freezing temperatures are a threat. But low temperatures are also rising because of the increased atmospheric levels of CO_2, methane, aerosols, and other things. Our "water bags" are favored by this. Agaves and cacti also welcome genetic intervention (Felker et al., 2009), as genes for low-temperature tolerance and especially low-temperature acclimation become better understood for all plant species.

Besides survival, temperatures also affect net CO_2 uptake and hence productivity by plants, topics covered in the next two chapters for agaves and cacti. Currently, corn (maize), rice, and wheat are cultivated in regions where the ambient temperatures are fairly optimal. Globally increasing temperatures will tend to reduce yields there for these major crops. Likewise, supra-optimal temperatures will also depress yields for agaves and cacti, which can be quantified based on their temperature responses of net CO_2 uptake presented in the next chapter (see Fig. 5-2).

Our CAM Plants Coping with Changing Rainfall

Things get a little murkier with respect to rainfall, because its future predictions are not as good as for temperature and for atmospheric CO_2 levels. But more rainfall, no problem, as agaves and cacti respond favorably to more soil water, as discussed in the next chapter. Less rainfall, no problem, as being CAM plants, agaves and cacti are extremely resilient to drought. Although an oversimplification, as the climate changes, dry regions will tend to become drier and to remain drier for longer periods. Wet regions will tend to become wetter and remain so for longer periods. With regard to specifics, the devil is in the details.

Locally decreasing rainfall in the future can have major ecological repercussions. For instance, during a 30-year period at the end of the 20th century, only about 16% of the years had enough rainfall for seedling establishment by *Agave deserti* and *Ferocactus acanthodes* in the northwestern Sonoran Desert of California (Nobel, 1988). Similarly, *F. acanthodes* and *Carnegia gigantea* are found only in regions of Arizona and California where at least 10% of the years have enough rainfall to support seedling establishment, indicating the crucial importance of this phase in the life of a plant. The establishment rate of these species will decrease with the predicted future decrease of rainfall in the Sonoran Desert, probably restricting the ranges of these species even more than the projected habitat loss due to housing and other developments.

Establishment of new plantations of agaves and cacti likewise will revolve around the availability of water during the crucial first year. An accumulation of water stored in the leaves of agaves and in the stems of cacti is required for these plants to manifest their extraordinary tolerance of droughts (Chapter 3). Thus new plantings must be done judiciously, either providing water, presumably by drip irrigation, or gambling that weather conditions will provide the necessary rainfall. But agaves and cacti—with their utilization of Crassulacean Acid Metabolism and other adaptations to conserve water or to shift it between cells—are well positioned to respond to the challenges of changing future rainfall patterns. This suitability is both with respect to survival and to net CO_2 uptake, which is discussed in the next chapter.

5

CO_2 Uptake, Environmental Productivity Index

Now that we have considered the influences of environmental factors on the survival of agaves and cacti, including the specific effects of global climate change, we turn to a consideration of the influences of environmental factors on their CO_2 uptake and hence on their productivity. This is no easy task, as the various factors can be changing simultaneously and can have many different effects on a process as complicated as photosynthesis (Chapter 2). We need to strike a balance between accounting for all of the influences of light, temperature, and soil water availability on net CO_2 uptake versus dealing with these variables in a practical, workable manner that emphasizes the main effects. To do this we will introduce an *Environmental Productivity Index*, which has the acronym EPI.

Acronyms such as CAM, WUE, and EPI save the repetitious use of many words, but they can be confusing and annoying. A cheese-loving friend of mine, when he comes to visit, sees the sign on my house "Casa de CAM" ("House of CAM"). He begins drooling and asks if I have any *Cam*embert cheese!

A parking lot attendant for a local bank gave me a receipt saying "EPI." Initially thinking he might know about my involvement with plant science, I soon learned that the acronym stood for Enterprise Parking Inc. "EPI" has also been used for the Economic Policy Institute (a think tank), an Environmental Performance Index (evaluating human impacts on the environment), the European Patent Institute, a stock on the New York Stock Exchange, various companies (e.g., Eldridge Products Inc., Electrochemical Products Inc., Energy Products of Idaho), and a disease of dogs and cats (Exocrine Pancreatic Insufficiency), among others. But here EPI stands for the Environmental Productivity Index, which will initially be used to assess the effects of light, temperature, and soil water status on plant CO_2 uptake.

EPI—We Must Multiply the Factors

We are next faced with what mathematical form to use when representing the various environmental factors influencing CO_2 uptake. We could choose an additive form, where the total CO_2 uptake expected is based on a summation of the contributions from each environmental factor—when all of the factors are optimal, the index has its maximal value. On the other hand, suppose that some environmental factor basically stops CO_2 uptake, such as prolonged drought that eliminates stomatal opening. It then does not matter what the light level and the temperature are, as no CO_2 uptake would occur. So an additive form is not appropriate.

On the other hand, complete limitation by a single factor can be mathematically presented by multiplying the factors, one for each environmental variable. Again the index has its maximal value when all of the factors are optimal, but now the actual limitations of each factor can be realistically presented. For instance, when some factor eliminates net CO_2 uptake, its individual index is then 0.00. When that factor is optimal for net CO_2 uptake, its individual index is 1.00. And when some factor halves net CO_2 uptake, its index is 0.50.

Let us now give an equation for the Environmental Productivity Index (Nobel, 2009):

$$EPI = \text{Light (PPF) Index} \times \text{Temperature Index} \times \text{Water Index}$$
$$(5.1)$$

As just indicated, each individual index ranges from 0.00, when that factor causes no net CO_2 uptake to occur, to 1.00, when that factor is optimal for net CO_2 uptake. Hence EPI (Equation 5.1) indicates the fraction of maximal net CO_2 uptake expected based on the influences of these three environmental variables. EPI times the *maximal* net CO_2 uptake, which is measured over a 24-hour period for CAM plants under optimal conditions, indicates the *actual* net CO_2 uptake expected over a 24-hour period. To make this statement more concrete, we begin with a consideration of light, which is obviously necessary for CO_2 uptake and photosynthesis.

Light Responses

Our first focus is on how light affects the daily gas exchange patterns of agaves and cacti (Fig. 2-4). But their stomatal opening and hence net CO_2 uptake occur primarily at night when there is no sunlight. How are we going to perform an experiment measuring the light responses of CAM plants? For C_3 and C_4 plants this is easy— we simply vary the light level while the plants are in a gas exchange apparatus (see Fig. 2-2 for a non-portable one) and measure the rate of net CO_2 uptake.

CAM plants also respond to light. But the CO_2 uptake at night is not responding to the instantaneous light level, which is then zero. Rather, CAM plants respond to the total amount of light received during the previous daytime (or a series of daytimes, if the light available varies considerably from day to day). Thus to get a single point on a light response curve for a CAM plant, we must monitor the net CO_2 uptake over an entire 24-hour period and compare it to the total amount of light received during the daytime.

To get started we have used the term "light," such as that provided by the sun or a light bulb. Such radiation is used to power photosynthesis, but different types of light cause different responses (recall the comment in Chapter 2 about the hapless fellows who used tungsten lamps, with most of the radiation in the infrared region not useful for photosynthesis, instead of fluorescent lamps to raise the illegal plants hidden in their basement). To adjust to what plants do, we need to count the discrete units of light, which are termed *photons*. Photons are absorbed by photosynthetic pigments, such as chlorophyll. The rate of photons incident per unit area of an agave leaf or a cactus stem is called the *photosynthetic photon flux density* or more simply the *photosynthetic photon flux* (PPF; often formerly called the photosynthetically active radiation).

To measure the response of a CAM plant to light, we add up all of the photons received over the course of a daytime. This total daily PPF is then correlated with the total amount of CO_2 taken up over a 24-hour period. A 24-hour period is required, because although most net CO_2 uptake for agaves and cacti occurs at night, some occurs during the daytime (Fig. 2-4). Such total daily net CO_2 uptake is directly related to the biomass productivity that we will consider in the next chapter.

Figure 5-1 summarizes the response of agaves and cacti to the total daily photosynthetic photon flux, as presented by the Light (PPF) Index. A total daily PPF of zero indicates that the plants are maintained continuously in the dark. No photosynthesis can then occur but respiration does (until the plants die), so a small amount of CO_2 is released (generally slightly less than 1% of the maximal rate of CO_2 uptake). The total daily PPF must be increased to above 3 mol m^{-2} day^{-1} in order for positive net CO_2 uptake to occur over a 24-hour period for most agaves and cacti (Fig. 5-1). Such a low light level is adequate for reading and for certain house plants and understory forest species. But most agaves and cacti require substantial amounts of light. Thus these beautiful plants will not survive in the low-light levels of dimly lit restaurants and dark cafes, where they are sometimes placed.

As the PPF is steadily increased above 3 mol m^{-2} day^{-1}, the daily net CO_2 uptake rate steadily increases until it becomes maximal at a total daily PPF of about 30 mol m^{-2} day^{-1} (Fig. 5-1). This latter PPF level represents a well lit surface. In particular, it corresponds approximately to the amount of photons received per unit area by an unshaded vertical surface facing east or west on sunny days at most latitudes in the summer (for comparison, a maximum PPF of nearly 70 mol m^{-2} day^{-1} occurs for an unshaded horizontal surface on a clear day with the sun passing directly overhead at noon; Nobel, 1988).

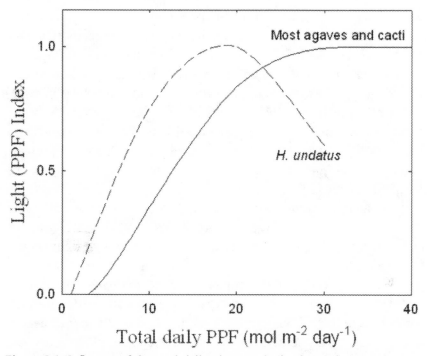

Figure 5-1. Influence of the total daily photosynthetic photon flux (PPF) on the Light (PPF) Index. Although the total daily net CO_2 uptake varies with plant species and environmental conditions, the shape of the light response curves for most agave and cactus species measured so far, including the four species considered here, *Agave deserti, Agave fourcroydes, Ferocactus acanthodes*, and *Opuntia ficus-indica*, are remarkably similar (solid line; Nobel, 1988). The fifth species considered, *Hylocereus undatus*, shows a decrease in total daily net CO_2 uptake at the higher PPF levels (dashed line; Raveh, Gersani, and Nobel, 1995; Nobel and De la Barrera, 2004). All plants were under wet conditions and approximately optimal temperatures; they were maintained for 8 to 12 days at a particular total daily PPF in order for the daily gas exchange patterns to stabilize.

Further increases in PPF above about 30 mol m^{-2} day^{-1} do not increase the net CO_2 uptake rate—we say that the photosynthetic processes are then *light saturated*. Interestingly, even agaves and cacti having very different maximal net CO_2 uptake rates, such as *Agave deserti* (Fig. 2-4A), *Agave tequilana* (Fig. 2-4B), and *Opuntia ficus-indica* (Fig. 2-4C), have almost identically shaped response curves for the Light (PPF) Index (Fig. 5-1).

But some cacti show a decrease in daily CO_2 uptake as the total daily PPF is raised above a moderate level of about 20 mol m^{-2}

day^{-1}. This is apparent for the hemiepiphytic cactus *Hylocereus undatus* (Fig. 5-1) raised all over the world for its fruits (Chapter 1). In its native habitats in Colombia and other countries, it occurs as an understory plant, where it does not experience full sunlight. Indeed, *H. undatus* is rather shade-tolerant and can have a net CO_2 uptake at a total daily PPF of only 1 mol m^{-2} day^{-1} (Fig. 5-1). When it is cultivated in Israel, it must be shaded by 20 to 60% to reduce the incoming light (mentioned in Chapter 1), as high light levels, such as full sunlight, not only reduce CO_2 uptake but also can cause permanent damage to the stems.

But the light responses of agaves and cacti pose another dilemma for predicting CO_2 uptake by whole plants. Namely, the leaves of agaves and the stems of cacti are opaque, meaning that light cannot pass through them. This is in marked contrast to the thin leaves of C_3 and C_4 plants, where light can pass through, allowing chloroplasts in leaf cells to intercept sunlight coming from either side of a leaf. For agaves and cacti, on the other hand, the side of a leaf or a stem facing away from the sun can receive a very low PPF even when the sun-facing side is light-saturated for photosynthesis. This low-PPF situation is especially apparent for the poleward-facing side of a vertical cactus stem in the winter.

Therefore we must consider the orientation of all photosynthetic surfaces to predict the total daily net CO_2 uptake of an entire agave or an entire cactus. Moreover, these plants require relatively high amounts of total daily PPF to saturate their total daily net CO_2 uptake (Fig. 5-1). Thus, limitations on net CO_2 uptake occur when one plant shades another, an important topic that we will return to in the next chapter when we discuss productivity.

Before leaving the topic of light responses, two morphological features related to the orientation of the photosynthetic surfaces of agaves and cacti are worthy of comment (Nobel, 1988). First, newly developing cladodes of opuntias tend to be oriented in a direction that maximizes light interception. At mid latitudes (for example, 30°N or 30°S), this generally means that new cladodes have a slight tendency to face east–west (if one side faces east, then the opposite side faces west), as this leads to the greatest light absorption over the course of a year. Likewise, when planting cladodes on flat ground, a slight photosynthetic advantage accrues to those cladodes facing east–west. However, for new cladodes developing near the winter solstice at mid latitudes or more poleward, there is a slight tendency to face north–south, as this leads to the greatest interception of light for the two surfaces combined at that time of the year.

Second, the leaves of agaves radiating from the base of a plant combined with the spacing angle of the leaves around the stem (to be discussed shortly with respect to the Fibonacci Series) leads to advantageous distribution of light over the surfaces of the leaves. A simple way to appreciate this is to imagine light absorption by an upside-down agave plant. The relatively horizontal leaves would then be on top getting full sun, which means they would be in the saturation zone for total daily net CO_2 uptake (Fig. 5-1). The central leaves would then be pointing downward and would be heavily shaded, so they would have very little net CO_2 uptake. The right-side-up orientation of an agave plant minimizes leaves that are light saturated and those having little or no net CO_2 uptake. Indeed, the more evenly that light is distributed over a plant, the higher the net CO_2 uptake by the whole plant.

Temperature Responses

Every biochemical reaction has an optimal temperature. A process representing a series of biochemical reactions, such as photosynthesis, also has an optimal temperature. With regard to net CO_2 uptake and plant productivity, such a temperature optimum is more realistic when considered over the portion of 24-hour periods when net CO_2 uptake is occurring (Fig. 5-2).

For agaves and cacti, the stomata open primarily at night, so most of the daily net CO_2 uptake occurs then (Fig. 2-4). Not surprisingly, the *nighttime* temperature has far more influence on the total daily net CO_2 uptake by such CAM plants than does the daytime temperature. Moreover, just as for tolerances of low temperatures or high temperatures (Chapter 3), the rates of net CO_2 uptake adjust or *acclimate* to new environmental temperatures. The Temperature Index (Equation 5.1) takes all of these factors into consideration.

For most agaves and cacti, the optimal average nighttime temperature is relatively low, about 14°C (57°F; Fig. 5-2). For such

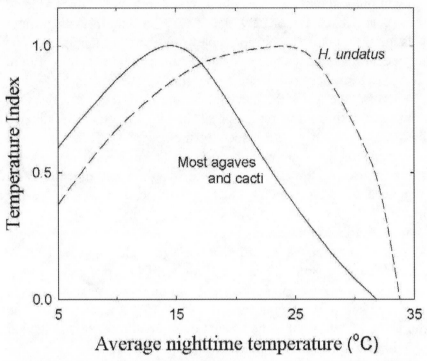

Average nighttime temperature (°C)

Figure 5-2. Influence of average nighttime temperature on the Temperature Index. The temperature response curves for most species of agaves and cacti measured so far, including the three species considered here, *Agave deserti*, *Ferocactus acanthodes*, and *Opuntia ficus-indica*, are remarkably similar (solid line; Nobel, 1988). The fourth species considered, *Hylocereus undatus* (dashed line; Raveh et al., 1995; Nobel and De la Barrera, 2004), has a higher optimal temperature. All plants were under wet conditions and a total daily PPF of about 20 mol m⁻² day⁻¹; they were allowed to acclimate to the day/night temperature regimes for 10 to 14 days before measurement.

plants, net CO_2 uptake is halved at average nighttime temperatures of about 4°C (39°F) and of about 23°C (73°F). Thus cool nights are favored for CO_2 uptake by these plants. Moreover, such cool temperatures also lead to high values of the Water-Use Efficiency (WUE; Equation 2.3). Recognition of the advantage of cool nighttime temperatures for maximizing net CO_2 uptake and also maximizing the WUE is crucial for the increased future cultivation of most agaves and cacti, such as *Opuntia ficus-indica*.

Analogous to its different response of net CO_2 uptake to PPF (Fig. 5-1), *Hylocereus undatus* also has a different response to temperature than most agaves and cacti (Fig. 5-2). In particular, this species native to tropical regions has no problem with warm nighttime temperatures. Its maximal daily net CO_2 uptake occurs for an average nighttime temperature of 25°C (77°F; Fig. 5-2). Warm nighttime temperatures favorable for its net CO_2 uptake and hence productivity occur in Colombia, Israel, Vietnam, and various other regions where it is profitably cultivated for its delicious fruits. In summary, the Temperature Index quantifies the influence of the crucial nighttime temperature on net CO_2 uptake for agaves and cacti, and it can vary with the particular species considered.

While working on EPI and the Temperature Index in May 1982 in Chile, my return flight on Braniff Airlines was supposed to go to Los Angeles, California. However, Braniff was going bankrupt, had cancelled the flight, but then reinstated a slightly later flight at the last minute in order to get the airplane itself back to the United States. Hence they had no weight limit. So we collected 200 kilograms (440 pounds) of cladodes of *O. ficus-indica*, which were loaded for free onto the plane. When we landed in Dallas, Texas, the plane ended its Braniff career, and I was stuck with all of the cladodes. The cladodes were promptly impounded by the United States Department of Agriculture, as my U.S. plant import permit was only for Los Angeles.

Somehow I got back to Los Angeles and called Washington, D.C., and explained the situation. To satisfy our international collaborative research commitments, I said with mock seriousness that the cladodes had to be maintained at 14°C at night (Fig. 5-2). The agent said sorry, we can't do that, so we will have to send all of the odd-looking plant segments to you. The cladodes soon arrived safely in Los Angeles at no charge. We then were able to do the necessary measurements for the Temperature Index and other aspects of EPI in controlled environment chambers using the same genetic material as for the field studies in Chile, whose productivity is considered in the next chapter.

Responses to Soil Water

The third component of the Environmental Productivity Index (EPI) as presented (Equation 5.1) is the Water Index, which quantifies the influence of soil water that can be taken up by the roots on the net CO_2 uptake by the shoots. Because the individual indices are multiplied together to obtain EPI, the order of the factors is immaterial. The order chosen here reflects the order of our consideration of the environmental variables, where we first considered the necessity of light for photosynthesis (Chapter 2). We next considered the effects of temperature and drought, primarily with respect to survival (Chapters 3 and 4).

The observations that led to the conceptualization of EPI began with *Agave deserti* in the Sonoran Desert of California in 1975. Not surprisingly, spurts of growth appeared to be associated with recent rainfall in this desert region. This crucial aspect of desert life emphasizes the importance of water, which when available in the soil removes the limitation on net CO_2 uptake and hence growth by this environmental factor. The Water Index (Equation 5.1) is thus 1.00 soon after rainfall. But droughts are inevitable in deserts.

Based on our technical definition of drought in Chapter 3, we can say that drought begins when the plants cannot take up water from the soil due to the respective values of hydrostatic and osmotic pressures in the soil versus the roots. Because roots can occur at many depths in the soil, this definition is ambiguous in a practical field sense. Thus the Water Index is here based on the soil water status versus the plant water status for roots in the center of the root zone. As a drought continues, the Water Index eventually falls to 0.00. Net CO_2 uptake over a 24-hour period by the plant is then abolished in response to the loss of its ability to take up water from the soil.

The effects of drought duration on total daily net CO_2 uptake vary greatly with species, even among CAM plants (Fig. 5-3). However, a guiding principle for understanding differences among plants is their relative water storage ability or *succulence*. Thus thick leaves of agaves or thick stems of cacti have a lot more water storage capacity per unit surface area than thin ones and hence can sustain photosynthesis for longer periods during drought (Fig. 5-3). Plant biologists tend to view this based on a volume per area paradigm—the greater the volume for water storage per unit surface area across which water can be lost, the greater is the drought tolerance. Of course, succulence is only part of the picture, as roles are played by (1) the tightness of stomatal closure during drought, (2) the thickness of the cuticle, and (3) the specific environmental conditions involved.

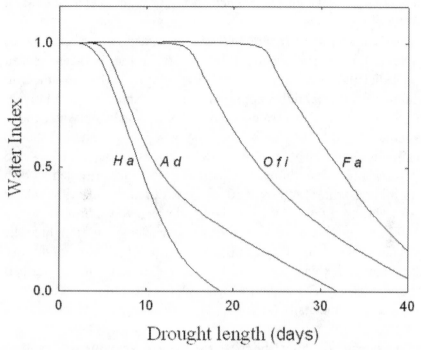

Drought length (days)

Figure 5-3. Influence of drought length on the Water Index for an agave and three cactus species differing in shoot succulence. Drought is defined as when the soil is too dry in the center of the root zone for the plants to take up water. The curves are for *Hylocereus undatus* (*Hu*; Raveh et al., 1995; Nobel and De la Barrera, 2002b), *Agave deserti* (*Ad*), *Opuntia ficus-indica* (*Ofi*), and *Ferocactus acanthodes* (*Fa*; Nobel, 1988). All plants were under wet conditions and a total daily PPF of about 20 mol m^{-2} day^{-1}.

Figure 5-3 indicates that the stems of *Hylocereus undatus* and the leaves of *A. deserti* can sustain a drought of about 10 days before the Water Index is reduced to 0.50, meaning that the net CO_2 uptake ability has been halved by the water stress. These stems and leaves do not store as much water per unit surface area as the approximately three-fold thicker cladodes of *Opuntia ficus-indica* or the barrel cactus *Ferocactus acanthodes* with its large diameter. These latter two species have a much slower decrease in total daily net CO_2 uptake during drought than the former two species (Fig. 5-3). Specifically, the Water Index is not halved until 23 days of drought for *O. ficus-indica* and until 32 days of drought for *F. acanthodes*. Thus the influence of morphology on the Water Index and hence on the net CO_2 uptake characteristics during drought is clearly evident for these four CAM species.

As a practical application of the Water Index, the response curves of net CO_2 uptake versus the soil water status (Fig. 5-3) can be used to predict the most efficient timing of irrigation. For instance, to maintain a substantial amount of net CO_2 uptake, the relatively precious commodity—water—can be applied once the Water Index is reduced to 0.50 instead of continually supplying excess water to maintain a higher value of this index. Because roots of agaves and cacti respond within hours to a rainfall or to water provided by drip irrigation, a Water Index of 0.50 can be restored to 1.00 within about half a day to a few days after the plants receive water. During the entire time interval from wet conditions to an ensuing drought to the halving of the Water Index to the restoration of the wet conditions, the Water Index can average 0.85 (see Fig. 5-3 for the drying part of this interval). Thus maximal net CO_2 uptake and productivity are reduced only modestly (about 15%) by this judicious application of water.

Besides indicating the best time for irrigation, the esoteric concept of EPI (Equation 5.1) can make someone a lot of money in other ways. This point was driven home to me by a farmer in Chile growing *O. ficus-indica*. He said that changing the traditional spacing and management for his crop to one based on EPI increased his annual profit by more than the price of a Mercedes autumbile!

Leaf Unfolding versus EPI

Because of the difficulty and expense of measuring CO_2 uptake in the field, it would be great to have a simple morphological surrogate to represent the effects of this process and its consequences for growth. One such idea occurred to me during an airplane flight to Merida, Yucatán, Mexico, in 1982. The trip was to study *Agave fourcroydes*, the agave used for centuries for its harvestable leaf fiber (Chapter 1). A technique to estimate productivity nondestructively in the field was highly desirable. My idea was to count the number of new leaves unfolding from the central spike of folded (unopened) leaves.

Perhaps determining the number of newly unfolding leaves for agaves would indicate the approximate biomass productivity, which hopefully could be predicted by the Environmental Productivity Index, EPI.

Successive agave leaves unfold at amazingly precise angular intervals around the stem of approximately 137°. Can this angular spacing be predicted? To do so, we consider Leonardo Fibonacci (1170–1250), an Italian mathematician reckoned by some to be the most talented one of the Middle Ages. He looked into the series (not actually discovered by him) formed by adding the previous two numbers to form the next number, namely, 0,1,1,2,3,5,8,13,21, etc. I like this Fibonacci Series so much that I have used it for combination locks and computer passwords, but here we will apply it to leaf unfolding for agaves.

How does the angular spacing of 137° relate to the Fibonacci Series? If you take the ratio of a number in the Fibonacci Series to the number occurring two more along the series (e.g., 3/8, 5/13, 8/21, etc.) and multiply by 360°, the number of degrees in a circle, the result converges on 137°! This mysterious property, when related to leaf arrangement around a stem, brings us to the concept of *phyllotaxy*. Phyllotaxy, the arrangement of leaves on a stem, can be quantified by counting the number of leaves in a spiral arrangement around a stem to lead to a leaf directly above an underlying one. This geometry is advantageous for light absorption because of the leaf placement in three-dimensional space. In particular, the ratios 3/8 (meaning 3 turns around the stem is accounted for by 8 new leaves) or 5/13 are adequate to account for the angular spacing of newly unfolding leaves of approximately 137° for agaves. All hail to Fibonacci!

Practically speaking, phyllotaxy allows an easy determination of the sequence of the stiff new leaves (see Fig. 1-1) unfolding and hence the total number of such new leaves for agaves. Moreover, the spiny tip of an agave leaf is dead, so clipping the tip of the last leaf unfolded causes no harm to the plant. When the plant is subsequently revisited, the newly unfolded leaves are easily counted. We are now armed with a simple morphological technique to estimate growth, and the EPI analysis based on net CO_2 uptake to quantify the environmental conditions. Let's see if they are correlated.

Agave deserti—Variation with Elevation

To see the effects of environment on leaf unfolding, *Agave deserti* was chosen for study along its entire elevational extent in the northwestern Sonoran Desert in California. Most sites were in the Philip L. Boyd Deep Canyon Desert Research Center, which is a large nature reserve in the University of California system encompassing 2,469 hectares (6,122 acres). Land adjacent to the Reserve is administered by the Bureau of Land Management, from which we also had permission to do research. A few months before he died in 1989, Philip Boyd, who had been the first mayor of Palm Springs, a Regent of the University of California, and the force behind the establishment of the Reserve, gave us encouragement for such studies on *A. deserti*, one of the dominant plants in the northwestern Sonoran Desert.

For this project, sites were to be chosen at approximately every 80-meter (260-feet) increase in elevation, which is easy to design in an office but often hard to do in the field. My field assistant Terry Hartsock and I established thirteen monospecific plots of *A. deserti* (mononospecific means that only *A. deserti* occurred, so the biomas productivity per unit ground area could be interpreted more easily, as is discussed in the next chapter). The plots were along California Scenic Highway 78, which goes from the Coachella Valley up into the Santa Rosa Mountains just south of Palm Desert, California. Elevations ranged from 320 meters (1,050 feet), which is the lowest elevation where *A. deserti* is locally common, up to 1,245 meters (4,085 feet), basically the local high-elevation limit of this species.

One of the sites was particularly hard to reach, requiring a long hike across rough terrain full of rattlesnakes. On our last trip we had spotted a shortcut near some houses on a local road off of Highway 78. As we took off on the shortcut, we noticed a suspicious black cable along the ground. We followed it for a short distance and stumbled upon three marijuana plants, which were tall and healthy, because the black cable was actually a hose bringing water to the desert. A short distance further we began measuring our plants, as usual. Terry then said "Park, don't move!" He wore these dark wrap-around sunglasses so I could not see his eyes, but he was uncharacteristically serious. He then said, "Someone has a shotgun pointed at us, and we must get out of here!" I said, "O.K., we will, and we will use our old pathway. But first we must finish the measurements!" Fieldwork can have unexpected dangers!

To get some feeling for the relation between EPI and the monthly rate of leaf unfolding, let us first consider data for *A. deserti* at an intensively studied intermediate elevation (Fig. 5-4). As seen in Figure 5-4A, the Water Index varied tremendously seasonally over the two-year study period. This region of the Sonoran Desert has a rainfall pattern that is annually bimodal, as rainfall tends to occur in the winter and again in the late summer/early autumn. Substantial rainfall causes the Water Index to be 1.00, meaning that plenty of soil water is then available to the plants of *A. deserti* (some such water is also stored on the succulent leaves of this species).

Another characteristic of the Sonoran Desert rainfall is its year-to-year variability—the Water Index was 0.00 for one month in the first year but for five months in the second year (Fig. 5-4A). Variations in EPI (Fig. 5-4B) largely followed variations in the Water Index in this desert setting. Actually, this is characteristic of deserts, as water is generally the most limiting factor for net CO_2 uptake there. More novel for the current application is that the easily measured monthly unfolding of new leaves was highly correlated with EPI (Fig. 5-4C). This was very encouraging.

Now let us examine leaf unfolding by *A. deserti* in the summer versus elevation and try to interpret the results using EPI. Low elevations are too hot in the summer for optimal CO_2 uptake, so the Temperature Index increases as we go up in elevation into cooler regions [9°C (16°F) overall from 320 m to 1,245 m]. Also, rainfall increases with elevation, as occurs in many other regions (here about a three-fold increase from 320 m to 1,245 m), leading to an increase in the Water Index with elevation. The Light (PPF) Index had relatively little change with elevation for the essentially flat sites chosen. Thus, we predicted better biomass productivity in the summer for *A. deserti* as we go up in elevation, consistent with the nearly five-fold increase in the late summertime leaf unfolding. Specifically, 59 leaves unfolded on 80 plants from July through September at 320 meters, increasing to 277 leaves on the same number of plants at 1,245 meters (Nobel and Hartsock, 1986).

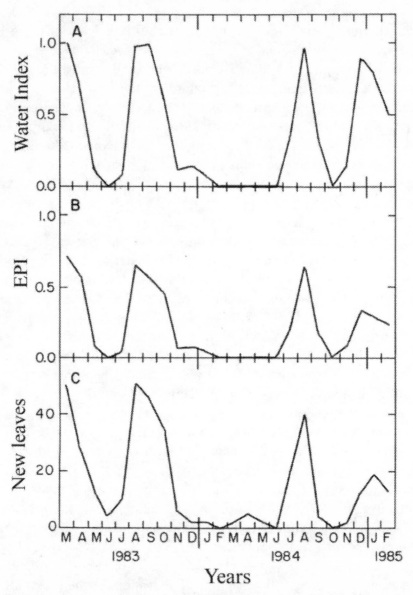

Figure 5-4. Monthly variation in various parameters for *Agave deserti* at an elevation of 840 meters in the northwestern Sonoran Desert in California over a two-year period: (A) Water Index, generally the most important influence on net CO_2 uptake and productivity in deserts; (B) Environmental Productivity Index (EPI), which is here controlled mainly by the Water Index but also takes into consideration local temperatures and light (PPF) levels; and (C) number of new leaves unfolding monthly on the 50 plants monitored, which is a proxy for the monthly biomass production by these plants. [Modified from Nobel and Hartsock (1986).]

What happens in the winter? Rainfall again increased three-fold with elevation, but temperatures then become too cold at the higher elevations for maximal net CO_2 uptake by *A. deserti*. Consistent with this, EPI and leaf unfolding in the winter are maximal at intermediate elevations. Also in the winter, leaf unfolding was over ten-fold higher on steeply south-facing slopes (angle of about 45°) than on steeply north-facing slopes (also an angle of 45°). This is consistent with the Light (PPF) Index, which in the winter was much higher (0.61) for south-facing slopes than for north-facing ones (0.04; Nobel and Hartsock, 1986). (This is also consistent with the placement of household windows at higher latitudes, which at such locations tend to be on the south-facing sides of houses in the Northern Hemisphere.)

The simple technique of determining leaf unfolding can help us relate plant growth and hence biomass productivity to environmental conditions, as quantified by EPI. Before we consider refinements to EPI, let us consider the leaf-unfolding responses of four other species of agaves and some analogous responses of two species of cacti as their environmental circumstances vary.

Other Agaves/Cactus Analogs

The protocol of measuring leaf unfolding monthly was readily done for *Agave deserti* in California (Fig. 5-4), but would require monthly trips to Mexico for *Agave fourcroydes* for which I had no research funds. Hence a deal was struck with some biochemists in Yucatán, who needed certain chemicals for their research. I would bring the chemicals if they would count the newly unfolding leaves monthly on the marked plants, clip their tips to avoid counting these leaves again, and repeat this month after month.

When I returned to Mexico with their chemicals, the bottles were so awkward that I repackaged the white crystals into large zip-lock plastic bags. My luggage also contained many small empty zip-lock bags for plant samples for nutrient analysis (considered in the next section). The customs official at the Yucatán Airport, upon opening my luggage and spotting large bags of white crystals and small empty bags, suspecting me of another kind of business! He blew a whistle, whereon gun-toting police unceremoniously hustled me into a backroom. I was blurting, "Yo scientifico" (loosely, "I am a scientist"), but it took an hour for my innocence to be accepted under such highly suspicious circumstances.

Getting back to the main story, the successful correlation of leaf unfolding with EPI demonstrated for *A. deserti* (Fig. 5-4) allows a testing of the various environmental factors on EPI for other agaves, such as those used for fiber or beverages (Nobel, 1988). *Agave fourcroydes* is grown in the Yucatán for the fiber in its leaves (Chapter 1) in a relatively wet region receiving 1,000 millimeters (39 inches) of rainfall annually. Irrigation can increase its Water Index 23% from 0.81 to 1.00, which causes a similar 24% increase in EPI and the monthly rate of leaf unfolding. Leaves of *Agave lechuguilla* are harvested from the wild in the Chihuahuan Desert and other semi-arid regions of Mexico. In a region with 380 millimeters (15 inches) of rainfall annually, weekly irrigation raises the Water Index, EPI, and the monthly rate of leaf unfolding each about 60%. For these two species, shading by 60% approximately halves the Light (PPF) Index and the monthly rate of leaf unfolding, further demonstrating the utility of EPI in predicting growth and productivity.

For *Agave salmiana*, which is used for pulque and mescal production (Chapter 1), monthly leaf unfolding is also highly correlated with monthly EPI. Similar observations occur for one- and three-year-old plants of *Agave tequilana* cultivated for tequila production. Nearly mature six-year-old plants of *A. tequilana* apparently have most leaf unfolding supported by current net CO_2 uptake, as represented by EPI. But some additional leaf unfolding is apparently supported by the carbohydrates accumulating in the stem, which will soon be harvested for tequila production. In any case, the simple technique of measuring leaf unfolding on a monthly basis gives strong support to EPI (Equation 5.1) as a predictor for monthly growth of agaves in the field.

Monitoring the monthly growth of cacti is more challenging morphologically than for agaves. However, analogous to the unfolding of new leaves for agaves, the monthly emergence of new *areoles* can be monitored in the field for certain cacti. Areoles, which technically are lateral buds, are the little projections on the stems of cacti that produce spines and glochids as well as fruits and even new stem segments. For the barrel cactus *Ferocactus acanthodes*, this morphological feature of new areoles is highly correlated with the month-to-month values of its EPI. Also, EPI predicted an annual stem growth for *F. acanthodes* of 8% compared with 9% determined by conventional morphological techniques (Nobel, 1988).

Opuntia ficus-indica does not have a convenient morphological trait to monitor growth. Its new cladodes are initiated at various times of the year, and the cladodes change in size based on the growth conditions and their water content. Thus a destructive and more labor-intensive technique is necessary, namely, the harvesting and drying of cladodes to determine their biomass. Again, monthly increases in biomass are highly correlated with monthly values of EPI determined for this species.

EPI thus allows a prediction of net CO_2 uptake and productivity under various environmental conditions, both current ones and those expected in the future. It is hard to overstate my enthusiasm both for EPI and for the verification of a simple morphological correlate useful for its evaluation in the field for agaves. The high price of modern portable gas-exchange equipment (Chapter 2) and the experience necessary to use these instruments properly can thus be circumvented. Namely, productivity can be estimated using relatively simple morphological measurements and calculations based on EPI in many regions of the world where the increased cultivation of agaves and cacti is predicted for the future (Chapter 7).

Nutrients—Too Many Variables

Soil nutrients also affect the rate of net CO_2 uptake by agaves and by cacti but in a very complex manner, just as for other plants. (Adding humor to this crucial but perhaps boring section on nutrients proved difficult. A colleague agreed that nutrients are inherently not very funny.) Almost all plants respond positively to the application of a fertilizer containing nitrogen (N). But extra high levels of soil N can cause damage to plants and decrease their productivity, so care must be taken with fertilizer application. A typical commercial fertilizer contains specific values of NPK, where P refers to phosphorus and K to potassium. These three elements plus calcium (Ca), magnesium (Mg), and sulfur (S) are termed *macronutrients*, because they are usually required by plants in high amounts. Elements required in small amounts, such as boron (B), chlorine (Cl), copper (Cu), and zinc (Zn), are referred to as *micronutrients* (about eight micronutrients have been demonstrated or are suspected to be necessary for agaves and cacti), and they also are essential for plant growth.

On the other hand, sodium (Na) of table salt fame is generally inhibitory to the growth of agaves and cacti. And the content of hydrogen (H) ions also affects plant performance, as quantified by the pH (an acidic soil has a pH of 6 or below, and an alkaline soil has a pH of 8 or above)—agaves and cacti do better in acidic soils. Taking all of these factors into consideration leads to a Nutrient Index that can be inserted as a multiplicative factor into EPI represented by Equation 5.1 (Nobel, 1989). We next consider semi-quantitatively how the most important of these soil elements affect net CO_2 uptake and hence productivity of agaves and cacti.

Let us consider two extremes: (1) agaves and cacti growing in a relatively infertile soil characteristic of deserts, and (2) those growing in an agricultural soil that has been heavily fertilized. Growth of both agaves and cacti will generally benefit from an increase in soil nitrogen. For instance, if the nitrogen level in either soil is raised by 50% by fertilizer application, their net CO_2 uptake and hence productivity will generally increase by 15 to 20% (Nobel, 1989). A similar 50% increase in phosphorus will increase net CO_2 uptake by about 10% for agaves and cacti for the infertile soil with a lesser affect for the more fertile soil. A 50% increase in potassium may increase their net CO_2 uptake by 5% for the infertile soil but have little effect for the fertile soil.

The effects of micronutrients on net CO_2 uptake by agaves and cacti are more variable, as many soils have sufficient quantities of most of these elements. However, some soils are deficient in micronutrients, such as for boron (B) in the Chihuahuan Desert of Mexico. Agaves and cacti tend not to be native to saline soils that contain high levels of NaCl. In this regard, irrigating various such species with a solution containing one-fifth of the salt content (mainly NaCl) of seawater generally inhibits growth by about 50%.

To help quantify the Nutrient Index, net CO_2 uptake can be measured over a 24-hour period in the field under wet conditions. An all-purpose fertilizer containing both macronutrients and micronutrients can then be applied to the same plants or to adjacent ones, and the daily net CO_2 uptake measured again for these plants under wet conditions. Such a fertilizer treatment effectively raises the Nutrient Index to 1.00. Comparing the two measurements of total daily net CO_2 uptake will give an indication of any nutrient limitations and provides a numerical value for the Nutrient Index. This time-consuming process can be simplified based on an elemental analysis of the field soil compared with the known effects of soil elements on plants in general (Epstein and Bloom, 2005) and on agaves and cacti in particular (Nobel, 1988, 1989). In any case, understanding the nutrient relations of plants is complicated. Much more research is needed for agaves and cacti. But the bottom line is that nitrogen in the soil is good and table salt is bad.

Atmospheric CO_2—Some Good News

We indicated in the previous chapter that plants benefit from the increasing atmospheric CO_2 levels by having higher rates of net CO_2 uptake. For CAM plants in general and agaves and cacti in particular, net CO_2 uptake and biomass productivity increase about 35% as the atmospheric CO_2 level is doubled from about 350 ppm to 700 ppm. This corresponds to an easy to calculate 1% increase in net CO_2 uptake ability per 10 ppm increase in the atmospheric CO_2 level. Moreover, this suggests a practical form for a CO_2 Index to be inserted into EPI, as represented by Equation 5.1. Namely, the CO_2 Index can be represented as follows:

$$CO_2 \text{ Index} = 1.00 + (0.01)(CO_2 \text{ level in ppm} - 350) \quad (5.2)$$

When the atmospheric CO_2 level is 400 ppm, the CO_2 Index in Equation 5.2 is then $1.00 + (0.01)(400 - 350)$ or 1.05. This means a 5% increase in net CO_2 uptake ability over the value appropriate for the atmospheric CO_2 level occurring in the 1990s (Fig. 4.1). If the atmospheric CO_2 level becomes 700 ppm, Equation 5.2 indicates that the CO_2 Index for agaves and cacti is then $1.00 + (0.01)(700 - 350)$ or 1.35. This corresponds to the observed 35% enhancement in the total daily net CO_2 uptake caused by an experimental doubling of the atmospheric CO_2 level. At even higher atmospheric CO_2 levels than 700 ppm, the CO_2 Index for agaves and cacti will eventually saturate, but so far this higher concentration range has not received much research attention.

Next we examine some of the nuances accompanying the increasing atmospheric CO_2 levels for CAM plants (Drennan and Nobel, 2000). In the previous chapter, we mentioned that the chlorenchyma becomes thicker, root systems expand, and shoot development occurs more rapidly under elevated atmospheric CO_2 levels. This is good news for net CO_2 uptake by CAM plants. Secondary effects also occur, such as slightly higher optimal temperatures at higher atmospheric CO_2 levels and a tolerance of slightly longer droughts. This is also good news for net CO_2 uptake by agaves and cacti under future climatic conditions.

The enhancement in net CO_2 uptake caused by a doubled atmospheric CO_2 level is generally maintained for at least a year by CAM species. This is contrary to many studies on C_3 species, where the enhancements usually decrease after a few months. Averaged over the ten or so species of agaves and cacti that have been studied, the daytime proportion of net CO_2 uptake approximately doubles from 10% to 20% of the total daily net CO_2 uptake as the atmospheric CO_2 level is experimentally doubled. This is a consequence of various factors, especially the higher Water-Use Efficiency under higher atmospheric CO_2 levels (Chapter 4).

A doubling of the atmospheric CO_2 level results in lighter (more bleached) colors for the leaves and the stems of CAM plants. In particular, the chlorophyll content is reduced about 15%, as the extra availability of CO_2 actually requires fewer photosynthetic pigments for processing the photons. Thus, increases in the atmospheric CO_2 level will lead to changes in the morphology and the anatomy of agaves and cacti, but the overall effects are positive and predictable as far as net CO_2 uptake and the CO_2 Index (Equation 5.2) are concerned.

Alas, a More Complex EPI

Including effects of both nutrients and increasing atmospheric CO_2 levels demands a more inclusive EPI than the one presented in Equation 5.1. This can be represented as follows:

$$\text{EPI} = \text{Light (PPF) Index} \times \text{Temperature Index} \times \text{Water Index}$$
$$\times \text{Nutrient Index} \times CO_2 \text{ Index} \qquad (5.3)$$

The Nutrient Index is rather complicated, as already indicated, whereas the CO_2 Index is quite straightforward (Equation 5.2). In any case, Equation 5.3 is ready to cope with the realities of different soils and the future changes in environmental conditions with respect to CO_2 uptake by agaves and cacti. This EPI times the maximal total daily net CO_2 uptake under ideal conditions (high light, optimal temperatures, wet soil full of nutrients) and the relevant atmospheric CO_2 level gives the actual daily CO_2 uptake expected under the specific environmental conditions considered.

Is such an Environmental Productivity Index (Equation 5.3) an oversimplification? Yes. Secondary interactions among factors occur, as already mentioned for the CO_2 Index. For instance, during drought the total daily PPF required to saturate total daily net CO_2 uptake decreases. However, as far as EPI is concerned, this slight increase in the Light (PPF) Index at most light levels is overwhelmed by the greater accompanying decrease in the Water Index.

The optimal temperature for net CO_2 uptake shifts downward as drought duration increases for plants in general and for agaves and cacti in particular. This downward shift reflects stomatal closure, which becomes fractionally greater at the higher temperatures. However, again in terms of EPI, the drought causes a much greater reduction in the Water Index than any accompanying changes in the Temperature Index. As the nighttime temperature deviates from the optimal one (Fig. 5-2), the total daily PPF required for light saturation decreases. However, such a lowering of the Temperature Index by 50% raises the Light (PPF) Index by less than 10%. In all cases, the proportionally greatest secondary interactions occur when EPI is already greatly reduced by one of the environmental factors, leading to only minor effects on the annual productivity to be discussed in the next two chapters.

In summary, EPI as represented by Equation 5.1 or 5.3 gets the job done. Namely, predicting the influence of soil and environmental factors in a way that is useful for estimating productivity in various regions and for various climates, both current and those expected in the future. In arid and semi-arid regions, the Water Index is usually the environmental factor most limiting to EPI (Fig. 5-4; Nobel, 2009). But the simplicity of EPI belies its beauty, especially when the details can be overwhelming and hinder decisions. Indeed, having a simple, physiologically based model that takes into consideration the most important consequences of environmental factors for net CO_2 uptake and hence productivity can help reveal the best course for action under new conditions.

6

Biomass Productivity—Tricks of the Trade

We have been considering net CO_2 uptake per unit leaf area of agaves and per unit stem area of cacti. This leads naturally to a consideration of net CO_2 uptake and hence biomass gain per unit *ground area*. *Biomass* refers to the solid material, usually mainly carbohydrates but excluding water, that accumulates in plants as a result of photosynthesis. Such a consideration is often more meaningful for the well being of an ecological community or the harvest of an agricultural field than is the CO_2 uptake by a shoot. For instance, an isolated plant unshaded by its neighbors will exhibit the greatest biomass gain per plant. However, the maximal biomass gain per unit ground area occurs when such plants are packed much more closely together. The trick is to get some but not too much interplant shading. We can evaluate this with respect to net CO_2 uptake using the Light (PPF) Index developed in the previous chapter, which relates the sunlight received by a leaf or a stem surface to its total daily net CO_2 uptake.

Units and Conventions

   Biomass productivity is generally expressed as tonnes of dry matter produced per hectare (10,000 m^2) per year, where 1 tonne hectare^{-1} equals 0.446 ton acre^{-1} (in Chapter 1, we indicated that 1 tonne is 1,000 kilograms or 2,205 pounds; 1 ton is 2,000 pounds). Because the water content of plants varies with the availability of water in the soil, such as decreasing during drought, the most reliable data for comparisons among plant species are not expressed on the basis of harvested weight, which is generally termed *fresh weight* (sometimes also *wet weight*). Although fresh weight is an easier quantity to measure, more useful data are expressed based on *dry weight*, such as that obtained after drying the harvested material at 80°C (176°F) until no further weight change occurs. Drying can take about 24 hours for thin pieces of the succulent leaves of agaves or the stems of cacti, but 3 or 4 days are needed to dry more massive plant parts.

Although relatively easy, such drying of plant material can be a nuisance. For instance, leaves of *Agave mapisaga* and *Agave salmiana* can weigh 40 kilograms (90 pounds) each. When new to the study of plants with such large leaves in the early 1980s, I instructed an assistant to harvest eight leaves in a range of sizes from each of eight randomly chosen plants of *A. salmiana* in the field, usually a modest request for plant ecophysiologists. When I checked on the progress of the drying, the smell was overwhelming. The harvested leaves had a fresh weight of over 1,000 kilograms. Every oven available in every laboratory of the whole Institute had been commandeered for the drying of the leaves. Obviously a gap can occur between a request from the Ivory Tower and what is really practical.

Besides expressing productivity on the basis of dry weight, which avoids the daily and the seasonal fluctuations in water content, another general practice is to report biomass productivity of the shoots only. Shoots are readily harvested, but biomass accumulation also occurs in the roots. Roots are time consuming to excavate and usually difficult to obtain in their entirety (some small roots can break off and remain in the soil, while other roots go very deeply into the soil). This is less of a problem for the total biomass productivity of agaves and cacti, as their roots generally contribute only 8 to 12% of the total plant biomass compared with a contribution of 30 to 50% for the roots of C_3 and C_4 perennials (mentioned in Chapter 3).

The lower biomass allocation to the roots for CAM plants is basically a consequence of their high Water-Use Efficiency (Chapter 2). In particular, less water uptake from the soil is necessary to keep the plants happy. This allocation pattern advantageously allows a greater fraction of the plant biomass to be harvestable as shoots for agaves and cacti compared with C_3 and C_4 perennials, as was also mentioned in Chapter 3.

Coaxing High Productivities Based on EPI

Increasing the productivity of *Agave tequilana, Opuntia ficus-indica,* and certain other species can require changes from the traditional plant spacing that has often been used for hundreds of years. Such changes can be based on the Light (PPF) Index (Fig. 5-1) of the Environmental Productivity Index (EPI; Equations 5.1 and 5.3). The paradigm is simple. Namely, any sunlight hitting the ground is mostly wasted as far as photosynthesis is concerned (a small fraction of the sunlight hitting the ground is reflected by the soil, which can be intercepted by plants and used for photosynthesis).

For agaves and cacti under cultivation, the rows are often too far apart to achieve maximal productivity per unit ground area. Admittedly, such wide spacing between rows of plants facilitates various agricultural practices, such as weeding, picking of fruits, and general plant maintenance. Also, wide spacing can be advantageous for a plant in obtaining water and nutrients from greater regions of the soil. And the land may not currently be worth very much, so saving ground area may not be a high priority, although agricultural land is becoming increasingly valuable. In any case, decisions must be made regarding the trade-offs between closer spacing of the plants, the accompanying reduction of the Light (PPF) Index, and the biomass productivity per unit ground area. The latter generally achieves its maximal value when the plants are quite close together.

Agave fourcroydes, Models, and Maya

Let us next consider the spacing and the productivity for *Agave fourcroydes* in the Yucatán peninsula of Mexico, where it is harvested for the fiber in its leaves (Chapter 1). A key parameter at the plant level is the *Leaf Area Index* (LAI), which is the total leaf area per unit ground area. The two sides for the leaves of *A. fourcroydes*

are considered separately, because the leaves are crescent-shaped
in cross-section. Hence the upper leaf surface has a slightly smaller
area than the lower leaf surface. Also, the opaqueness of the leaves
forces an analysis based on the different orientations of the two sides
of a leaf, as light interception depends on the direction of sunlight
versus the direction faced by a particular leaf surface (obviously
different for the two sides of these opaque leaves).

As mentioned in the previous chapter, the technique of using
leaf unfolding to estimate productivity was first developed for *A.
fourcroydes*. Armed with EPI and computer models, the optimal
spacing for this species was examined. The computer model was
developed for a large plant with 160 leaves radiating from its base.
For modeling purposes, each leaf was initially subdivided into eleven
surfaces with different orientations to estimate how much light was
absorbed by the leaf. As the model became more sophisticated and
computing became cheaper, each leaf was later subdivided into over
1,000 small surfaces. Having many subdivisions of a leaf's surface
area allows the effect of shadows, which may occur on only part of
a leaf and markedly affect photosynthesis locally, to be represented
more accurately.

As the plants are moved closer and closer together in the
computer model, more interplant shading occurs. The local Light
(PPF) Index (Fig. 5-1), which represents the fraction of maximal net
CO_2 uptake due to the light incident per unit leaf surface area, then
decreases. Although the shading increases, the productivity per unit
ground area initially increases as the plants get closer and closer
together and more of the light is intercepted by them (Nobel and
Garcia de Cortázar, 1987; Nobel, 1988). Biomass productivity of
A. fourcroydes was also checked using the monthly leaf unfolding
(e.g., Fig. 5-4 for *Agave deserti*) at specific values of the Leaf Area
Index occurring in the field.

Both the model and the measured rate of leaf unfolding indicated
that the productivity was maximal at a Leaf Area Index of about
6, meaning that the total area of both sides of the leaves of these
agaves was then six times the ground area underlying the plants. As
the plants were moved even closer together than for an LAI of 6,
which was relatively easy to do in the computer model but hard to

do experimentally in the field, the productivity per unit ground area began to decrease. Too much shading then occurred, so a greater fraction of the total leaf surface area had a net release of CO_2 (Fig. 5-1) when the plants were very close together.

Remarkably, the optimal spacing of *A. fourcroydes* deduced by the computer model was within 2% of the spacing developed empirically many, many centuries earlier by the Maya, who also harvested the leaves for their fiber content. But at least an interpretation based on light interception and net CO_2 uptake was now available. A similar computer model was developed for *Agave tequilana*, which has broader and thicker leaves than does *A. fourcroydes*. The optimal Leaf Area Index for *A. tequilana* for biomass productivity per unit ground area is also about 6 (Nobel, 1988). This requires a much closer spacing between rows of plants than that used traditionally, a change that has recently been adopted for some high-productivity plantations in Jalisco, Mexico.

Opuntia ficus-indica, Models, and EPI

A modeling-plus-measurement approach to study optimal plant spacing has also been done for *Opuntia ficus-indica*. Instead of the Leaf Area Index to express the total photosynthetic surface area per unit ground area, the *Stem Area Index* (SAI) is used for this cactus. SAI is the total stem surface area, including the areas of ribs for barrel and columnar cacti, per unit ground area.

First, a word about the computer model, which incorporates the sizes and the shapes of cladodes. A cladode (or "pad") is like a large waffle or a large and very thick pancake; it is often oval or nearly circular (Fig. 1-1). The two sides of a cladode and its peripheral margin (the circumscribing edge) are represented in the model by 300 subdivisions of a cladode's surface area. Again, many subdivisions are needed so that the effects of shadows cast by other cladodes on the same plant or shadows cast by adjacent plants can be represented accurately with respect to net CO_2 uptake. The size and the orientation in space of each of 128 cladodes on a rather large specimen of *O. ficus-indica* were carefully determined in the field in order to create a realistic three-dimensional representation of all of the photosynthetic surfaces for this particular plant.

Next this plant was bombarded in the computer model with 100,000 parallel beams of sunlight. The angles of the sunlight were varied hourly throughout the daytime. Diffuse radiation from the sky and from clouds was also simulated. The hardest part was calculating the shadows cast by one cladode on another. This became more important as the plants, which were actually replicas of the one with 128 cladodes, were placed closer and closer together in the computer model. The total daily PPF was then calculated for the 300 small surfaces per cladode and the 128 cladodes per plant, or 38,400 small surfaces on each plant. Next the model used the Light (PPF) Index for *O. ficus-indica* (Fig. 5-1) to calculate the relative net CO_2 uptake for each such small surface over the course of a day.

Three different days with typical cloudiness were chosen—one near the summer solstice, one near the winter solstice, and one near an equinox (spring and autumnal). This led to the prediction of the optimal Stem Area Index for *O. ficus-index* (Nobel, 1988). If you guessed 6, you are right! To achieve an SAI of 6 for *O. ficus-indica* requires placing the plants closer together than the traditional planting practices for this cactus, especially when it is cultivated for fruit. Suggesting changes for plant spacing based on a computer model stirred up a hornet's nest of protest among some traditionalists. But certain entrepreneurs concerned with obtaining high biomass yield per unit ground area listened and were subsequently rewarded with much greater annual biomass yields in their fields.

While working on the model for *O. ficus-indica*, Victor Garcia de Cortázar set up three computers in the dining room of the three-bedroom apartment (flat) owned by the Universidad de Chile in which I was staying in Santiago. Due to political unrest, one night the police established a curfew, so he simply stayed all night tending the computers working away on the model. Other incidents also occurred in the apartment. One morning I awoke to find a caged rat in one of the bedrooms. Not any rat, but one with its own passport. The Brazilian scientist with the rat informed me that this special animal was involved in important medical research. Another night while just falling asleep in the largest of the three bedrooms, my door was suddenly thrust open by a man from Argentina. He looked surprised and then settled into one of the other bedrooms. A few hours later my bedroom door was again thrust open, but this time by a lady scientist from Spain. She began undressing until she realized the occupant of the bed had a moustache and was not the man she expected!

Once the best spacing was established for *O. ficus-indica* based on its Stem Area Index, emphasis shifted from light (PPF) to all of the other factors involved in EPI (Equation 5.3) to try to maximize biomass productivity. In Chile a planting site was chosen in an abandoned agricultural field that had been heavily fertilized; it was also near an irrigation ditch that leaked profusely. In Mexico, the plants were watered and fertilized weekly. Presto, at both sites the Water Index became 1.00 over the course of a year, and the Nutrient Index was maximized (see Equation 5.3).

Varying the Temperature Index is not practical under field conditions. It can mean moving the plants to a different site or artificially heating or cooling them, all of which are prohibitively expensive options. Thus to evaluate various possible regions for the cultivation of agaves and cacti, the Temperature Index (Fig. 5-2) is handy to quantify local environmental effects. Of course the nighttime temperature is key for net CO_2 uptake by these plants.

Indeed, cool temperatures at night are preferred for net CO_2 uptake by most CAM plants. The optimal temperatures are near 14°C (57°F) for the majority of agaves and cacti, including *O. ficus-indica* (Fig. 5-2). For the sites chosen in Chile and Mexico, its Temperature Index averaged 0.91, a relatively high value reflecting the relatively cool nights for most of the year at both locations. The plants flourished, and harvesting a portion of them after a year was eagerly awaited. The harvest showed that the biomass productivities were a remarkable 47 to 50 tonnes of dry weight hectare^{-1} year^{-1} (22 to 23 tons acre^{-1} year^{-1} (Garcia de Cortázar and Nobel, 1992; Nobel, García-Moya, and Quero, 1992). Why this is a remarkably high productivity will be dealt with in the next section. There the maximal productivities among the most productive plant species from all of the three photosynthetic pathways are examined, and comments are made on their biochemical underpinnings.

Optimal Leaf or Stem Area Index

The Leaf or Stem Area Index can be measured. And optimizing the plant spacing to lead to values near 6 is the most economical way to maximize biomass productivity of agaves and cacti per unit ground area under most field conditions. In this regard, the area of an agave leaf is approximately its width at midleaf times its length times two (to include both sides) times about 1.1 (to account for their crescent shape, which can readily be checked for a particular species). The area of an opuntia cladode is approximately its width times its length times two (to include both sides) times about 1.05 (to account for the peripheral margin of these relatively thick organs) times $\pi/4$ [0.785, the same factor relating the area of a circle to its diameter squared, namely, a circle's area equals π (diameter)^2/4].

Land covered by plants typically has a Leaf Area Index of just over 4 in many ecosystems (Nobel, 2009). Such an LAI ensures that most of the incoming sunlight is absorbed and that very little of it hits the ground and hence is basically wasted as far as photosynthesis is concerned. For the thin leaves of C_3 and C_4 plants, by convention this Leaf Area Index is based on one side of the leaves, instead of both sides, as is used for the thick and opaque leaves of agaves. Thus the optimal Leaf or Stem Area Index with regard to biomass productivity of about 6 for agaves and cacti is essentially consistent with such an LAI near 4 based on one side (meaning 8 if both leaf sides are included) for the thin leaves of C_3 and C_4 plants. In both cases, the ground is advantageously covered (shaded) with respect to light absorption by the plants.

How to dance with the data and end up with the best partner for plant spacing? Happily, the biomass productivity of agaves and cacti is relatively insensitive to changes in the photosynthetic surface area per unit ground area near the optimal values. Specifically, compared with the optimal productivity at a Leaf or Stem Area Index of 6, biomass productivity at a value of 4 is reduced only 10% from the maximum and at a value of 8 is reduced only 2% (Nobel, 1988). Thus there is great flexibility in choosing the plant spacing that maximizes productivity, as lower net CO_2 uptake per plant is partially offset by greater net CO_2 uptake per unit ground area as the plants get closer and closer together above an LAI or an SAI of 4.

The Maya got the spacing right for harvesting maximal leaf fiber from *Agave fourcroydes*, even without computers and EPI to guide them. But *Agave tequilana* and *Opuntia ficus-indica* have traditionally been cultivated at spacings between rows that allow much sunlight to hit the ground, as anyone can observe upon viewing the beautiful plantations of these species in Mexico and around the world (Nobel, 1994). Wide spacing of course facilitates the harvesting of agave piñas and cactus fruits. But future applications designed to maximize biomass yield per unit ground area will require closer plant spacings than those used traditionally.

Data, such as the optimal LAI and SAI, are crucial for scientists and for planting adjustments based on global climate change. Speaking of data, once I was working on the thermal tolerances of cacti at the Nevada Test Site (where numerous below-ground detonations of atomic weapons have occurred, leaving pronounced undulations in the roads). I was staying in a spooky, army-issue, grey-metal trailer. Accidentally, they had issued a duplicate key to another, who entered my room in the middle of the night. Alarmed, I leapt out of bed and placed my body over my research notebook. Another time I was in a plane in the Mexico City airport; somehow the plane caught on fire during refueling. The loudspeaker was saying (in Spanish) to abandon your luggage and take off your high-heeled shoes. Without hesitation, I grabbed my data and placed it next to my body under my undershirt before sliding down the evacuation shoot! Scientists instinctively want to protect their data.

Highest Plant Productivities—Pathway Analysis

The "World Cup" to achieve the highest plant biomass productivity was very competitive among scientists in the latter part of the 20th century. Entries came from all three photosynthetic pathways. The competition from many countries promoting their various species played out over many years in numerous research articles and scientific books. The results are impressive. The most productive species from the C_3, the C_4, and the CAM pathways all can have an annual biomass productivity of 40 tonnes of dry weight hectare^{-1} year^{-1} (18 tons acre^{-1} year^{-1}; Table 6-1).

Table 6-1. The five highest, above-ground, annual, dry-weight productivities for each of the three photosynthetic pathways. Data were obtained up to 1990, when such measurements were relatively common, and are from Nobel (1991).

Pathway	Species	Country	Productivity (tonnes hectare^{-1} year^{-1})
C_3	*Cryptomeria japonica*	Japan	44
	Elaeis guineensis	Malaysia	40
	Eucalyptus globulus	Portugal	40
	Eucalyptus grandis	South Africa	41
	Pinus radiata	New Zealand	38
C_4	*Cynodon plectostachyus*	United States	37
	Pennisetum purpureum	El Salvador, U.S.	70
	Saccharum officinarum	Australia, U.S.	64–67
	Sorghum bicolor	United States	47
	Zea mays	Italy, U.S.	36–40
CAM	*Agave mapisaga*	Mexico	38
	Agave salmiana	Mexico	42
	Ananas comosus	United States	35
	Opuntia amyclea	Mexico	45
	Opuntia ficus-indica	Chile, Mexico	47–50

Perhaps the competition was not fair, as most of the C_3 winners were trees producing in their prime years. And the C_4 entries were mainly carefully bred crops. Very few CAM species entered the contest, although a few had an unfair advantage based on EPI. Productivity in tonnes hectare^{-1} year^{-1} for the five exceptional species in each entry class, all of which deserve a sporting congratulatory "High Five," average 41 for the C_3 plants, 52 for the C_4 plants, and 42 for the CAM plants (Table 6-1). This is not bad for agaves and cacti, especially when the most highly paid CAM plant, pineapple (*Ananas comosus*), substantially pulled down their average. There really is not much to dispute in these figures—they are measurements done relatively easily in the field. The data for the C_3 species are very consistent, but there is no reason to doubt the averages for species from the other two photosynthetic pathways.

Perhaps of more interest than the actual productivities is the question, "Can the relatively good performance of the CAM plants be explained?" After all, cacti have been dismissed as intriguing plants with picturesque shapes and beautiful flowers that grow slowly in deserts. How in the world can *Opuntia ficus-indica* exceed the biomass productivity of all but a few of the highest producing C_3 and C_4 plants in the whole world (Table 6-1)? To explore the theoretical basis for this, we need to delve a little deeper into the biochemistry of the three photosynthetic pathways (presented in Chapter 2). The key is how many energy-rich molecules are needed to support the biochemical reactions. Such molecules are called *energy currencies*, because a plant must spend them in order to obtain the products that it wants. As you might expect, we need to go into a few biochemical details to understand the relative productivity expected among the three photosynthetic pathways.

The biochemical process of photosynthesis is expressed by Equation 2.1 as follows:

$$CO_2 + H_2O \leftrightarrow CH_2O + O_2$$

where CH_2O represents a carbohydrate. The different cellular compartmentations of the three photosynthetic pathways are presented in Figure 2-1, indicating that C_4 and CAM plants are anatomically more complicated than are C_3 plants with regard to the fixing of CO_2 into carbohydrates. Next we present the energetics of the processes, meaning the minimum number of energy-currency molecules involved and hence the relative energy needed for CO_2 fixation in each of the three cases.

The energy currencies that support biochemical reactions occur as two main types: (1) adenosine triphosphate (ATP), which is famous as the molecule that leads to muscular contraction; and (2) a very electron negative or *reducing* compound that supplies a lot of energy in the presence of the relatively high levels of oxygen that are found in nearly all plants and animals. For photosynthesis, this latter energy currency is the reduced form of nicotinamide adenine dinucleotide phosphate (NADPH). Thus an energetic comparison of the fixation of a molecule of CO_2 by the three photosynthetic pathways comes down to the questions of how many molecules of ATP and how many molecules of NADPH are required.

The requirements in terms of how much of the two molecular energy currencies are needed for each of the three photosynthetic pathways are known. Each pathway requires two molecules of NADPH per molecule of CO_2 fixed, so no advantage to any pathway yet. The ATP requirement is variable. The C_3 pathway requires the least, three molecules of ATP per CO_2 fixed for the photosynthetic reaction that is summarized by Equation 2.1 (see just above). The three known C_4 pathways require more ATP per CO_2, four or five molecules. The CAM pathway requires even more, an average of six molecules of ATP per molecule of CO_2 fixed. Much of the extra ATP required for the latter two pathways is needed to concentrate CO_2 near Rubisco, which was alluded to in Chapter 2 and was emphasized in Chapter 4. We next consider the costs and the consequences of such raising of the CO_2 level near Rubisco with regard to the two reactions supported by this enzyme. This means that we must reckon with the cruel process of photorespiration, which releases CO_2 in plants in the light. But raising the CO_2 level near Rubisco reduces such photorespiration, so the extra ATP costs of C_4 and CAM plants provide those plants with a benefit.

Rubisco has the full name of ribulose-1,5-bisphosphate carboxylase/oxygenase (mentioned at the beginning of Chapter 2). The latter part indicates that it supports two different biochemical processes, CO_2 fixation (carboxylase) and CO_2 release (oxygenase). We indicated in Chapter 2 that the carboxylase is the good guy, leading to CO_2 fixation in photosynthesis. And that the oxygenase is the bad guy, leading to CO_2 release in photorespiration, which undoes the hard work of photosynthesis. Raising the CO_2 level at Rubisco, as for C_4 and CAM plants, favors the good guy. Thus C_4 and CAM plants have very little of the reputedly wasteful release of CO_2 by photorespiration.

But there is still some more bad news concerning the oxygenase activity of Rubisco, as it requires an additional input of energy. In particular, the many biochemical steps for photorespiration involve three different organelles (chloroplasts, peroxisomes, and mitochondria). Moreover, photorespiration involves some unusual chemical compounds and some energy-requiring reactions. Also, photorespiration increases more rapidly than does photosynthesis as the temperatures increases, so photorespiration becomes more important at high temperatures, such as above 30°C (86°F), and higher temperatures are associated with global warming. This subtle response of the key enzyme involved in net CO_2 uptake is part of the complexities of predicting plant responses to future climates.

In any case, C_4 and CAM plants have ingenious, if biochemically complex, ways to raise the CO_2 concentration near Rubisco (Fig. 2.1). This greater biochemical complexity results in four to six molecules of ATP being required per CO_2 fixed in C_4 and CAM plants compared with only three molecules of ATP for C_3 plants. Although there is a cost paid in the energy currency ATP, this local concentrating aspect for CO_2 favors the carboxylase activity of Rubisco. Favoring the carboxylase lowers the competing oxygenase activity of photorespiration and its CO_2 release by Rubisco as well as its extra energy requirements (Nobel, 2009).

When all of the ATP and the NADPH costs are considered and photorespiration is factored in, the C_4 pathway requires the least amount of energy per *net* CO_2 fixed (photosynthesis minus photorespiration). This is consistent with C_4 plants having the highest annual biomass productivity among the three photosynthetic pathways (Table 6-1). The CAM pathway can actually be as efficient or more efficient than the C_3 pathway, depending on nighttime versus daytime temperatures, among other considerations (Nobel, 1991). Thus, certain CAM plants equaling or outperforming C_3 plants has a sound theoretical biochemical basis, so their relatively high possible biomass productivity is no fluke. Indeed, this is a good omen for the future.

Productivity Revisited—More Realistic Values

We next pull back from the record-setting productivities under essentially ideal conditions and examine more realistic values. These are actually of greater importance in the big picture for agaves and cacti under typical field conditions, both now and in the future. Indeed, we all like Olympic records, but daily life is generally much more mundane, as are the usual productivities of agaves and cacti compared with the record-setting values (Table 6-1).

In deserts, the key for productivity is rainfall. For *Opuntia ficus-indica* in Tunisia, a northern Africa country where it is extensively cultivated for fodder in arid and semi-arid regions (Chapter 1), biomass productivity can be 2 tonnes of dry weight hectare^{-1} year^{-1} (1 ton acre^{-1} year^{-1}) at an extremely low annual rainfall of 100 millimeters (4 inches), 5 tonnes hectare^{-1} year^{-1} at 300 millimeters (12 inches), and 11 tonnes hectare^{-1} year^{-1} at 500 millimeters (20 inches) of annual rainfall (Nefzaoui and Ben Salem, 2002). These low biomass productivities are actually good news, as very few C$_3$ or C$_4$ perennials would even survive at 300 millimeters or less of annual rainfall, yet alone be palatable to animals.

The relevant biological parameter for biomass productivity in arid and semi-arid regions is not the high productivity possible under unlimited water conditions. Rather, it is the substantial productivity under more typical rainfall amounts plus the great Water-Use Efficiency (Chapter 2) of agaves and cacti compared with alternative C$_3$ or C$_4$ perennial crops. This aspect begs to be exploited more (considered in the next chapter). Optimal spacing based on EPI (Equations 5.1 and 5.3) was not incorporated into the Tunisian study just cited, so the productivities for *O. ficus-indica* could have been higher with closer spacing. In any case, agaves and cacti can be relatively highly productive under extremely adverse conditions of water availability.

Agaves and cacti can also be relatively productive in relatively infertile soils. For instance, *Agave deserti* is one of the dominant species at a series of low hills collectively referred to as "Agave Hill" in the University of California Philip L. Boyd Deep Canyon Desert Research Center in the northwestern Sonoran Desert, a Reserve that was mentioned in the previous chapter. In the monospecific (containing only one species) stands where its leaf unfolding was monitored (Fig. 5-4), it can have an annual biomass productivity of 7 tonnes of dry weight hectare^{-1} year^{-1} (3 tons acre^{-1} year^{-1}; Nobel, 1988). This is by no means a record yield, but let us next consider the circumstances.

The annual rainfall at Agave Hill then was 430 millimeters (17 inches), plenty of sunlight was available, and the nighttime temperatures were acceptably cool. But the soil had low levels of nitrogen (N; about 20% of typical agricultural soils), low levels of phosphorus (P; about 30% of typical agricultural soils), and low levels of potassium (K; about 60% of typical agricultural soils; Nobel, 1989; Dubeux et al., 2006). Appropriate fertilization with these three macronutrients will double the Nutrient Index (Equation 5.3) *of A. deserti* at Agave Hill and hence will double its annual biomass productivity. Nevertheless, even 7 tonnes hectare^{-1} year^{-1} is a respectable biomass productivity for relatively infertile desert soils that are low in nitrogen. Indeed, CAM plants in general and agaves and cacti in particular can overcome agriculturally poor conditions, both with regard to rainfall and to soil nutrients.

Because of ongoing ecological research, the Reserve is protected in places with chain-link fencing, including a gate for the gravel road leading to Agave Hill. One night someone pulled up the gate posts, apparently to gain access to the Reserve. The local sheriff came by the next day and asked if I drank a certain beer famous for the source of its water. I said no, I preferred such-and-such a brand. "Good," he said, as empty cans of the former beer had been found near the pulled-up gate posts. While working alone later that day, I heard a motorcycle approaching along the gravel road. The motorcycle was ridden by a big burly guy with a hunting rifle slung over his shoulder. Without thinking, I blocked his path and demanded to know what he was doing on the Reserve. He said, "Huntin' big-horns." I bravely retorted "This is a nature reserve, it is illegal to hunt here, and that big-horn sheep are protected." I told him to turn around and get out!

Which surprisingly he did. I went back to my field work shaking all over when it dawned on me what this guy could have done to me alone in the field.

To predict the biomass productivity of agaves and cacti in both new and old regions, we need to consider all of the environmental factors. The currently increasing atmospheric CO_2 levels are good news for net CO_2 uptake by them (Chapters 4 and 5). Soil nutrients are always difficult to evaluate. But the responses of agaves and cacti to NPK fertilization and to micronutrients (Chapter 5) are similar to the responses of other plants. Indeed, application of nitrogen to them may be cost effective in many situations.

Further decision making with regard to future increased cultivation of agaves and cacti often comes down to light, temperature, and soil water status (Equation 5.1). In this regard, the average nighttime temperature and hence the Temperature Index is site specific but readily quantified (Fig. 5-2). Knowledge is always better than ignorance, but moving to another location to plant agaves and cacti with better nighttime temperatures may not be a viable option. Increasing the capture of light means placing the plants closer and closer together, which is good up to a point (a Leaf or Stem Area Index of about 6). We must also consider contrary effects of close spacing on management practices. Raising the Water Index is basically a question of the economics of irrigation. And water is expensive to apply in many regions.

Sorting out all of the environmental factors affecting the biomass productivity of agaves and cacti is a challenge, but the rewards can be impressive. For instance, we can recognize the substantial productivity of these plants in regions unfavorable for the cultivation of C_3 and C_4 plants. At the beginning of this section we discussed the effect of rainfall on the biomass productivity of *O. ficus-indica* in northern Africa. Very few other species would be equally as productive under such arid and semi-arid conditions. Let us now consider various environmental factors on a larger scale for this species, which is already the most extensively cultivated CAM plant worldwide.

Under dryland farming (no irrigation) in central Chile with minimal fertilization, the annual biomass productivity of *O. ficus-*

indica can be 20 tonnes hectare^{-1} year^{-1} (9 tons acre^{-1} year^{-1}; Garcia de Cortázar and Nobel, 1990). Its Temperature Index in coastal central Chile is 0.95, but generally it is much lower at other locations. For instance, its Temperature Index averages 0.67 for North America and 0.69 for Africa. The Light (PPF) Index was based on a close spacing of the plants with a Stem Area Index of about 5, which is near the optimal for its biomass productivity. However, at the same spacing the Light Index averages 30% lower in the tropics due to clouds and 30% lower above 50°N or below 50°S than at 33°S in central Chile due to low sun angles in the sky during the winter. The Water Index averaged 0.51 in coastal central Chile, which is typical for frost-free regions of the Earth; but it can be much lower in deserts, such as the great expanses of the Saharan Desert in north-central Africa. Taking all of these factors into consideration, the worldwide productivity of *O. ficus-indica* is predicted to average about 12 tonnes dry weight hectare^{-1} year^{-1} (5 tons acre^{-1} year^{-1}) in frost-free regions (Garcia de Cortázar and Nobel, 1990).

A worldwide biomass productivity of 12 tonnes dry weight hectare^{-1} year^{-1} for *O. ficus-indica* is much less than the high measured productivities of 47 to 50 tonnes hectare^{-1} year^{-1} (22 to 23 tons acre^{-1} year^{-1}) under ideal conditions mentioned earlier in this chapter (Table 6-1). But life and plant productivity are rarely ideal. The biomass productivity of 20 tonnes dry weight hectare^{-1} year^{-1} in a region of Chile long used for opuntia cultivation is also more realistic than the maximum values. Moreover, *O. ficus-indica* also can have a biomass productivity of about 20 tonnes dry weight hectare^{-1} year^{-1} in northeastern Brazil, where it is extensively cultivated for fodder (Dubeux et al., 2006), as can *Opuntia ellisiana* in Texas (Han and Felker, 1997). We must use such more realistic productivities to evaluate future productivity of agaves and cacti, as is done in the next chapter. That chapter is the most crucial one in the book for interpreting their future in light of changing climates.

The high productivity of various opuntias was obvious at one level (former rampant growth of opuntias in Australia, outstanding yield in Chile and other regions) but was challenged in the scientific literature. No one likes to admit that a humble cactus can out-produce their favorite local crop for annual biomass yield. The biochemical justification presented above convinced certain critics of the high potential productivity of certain CAM species. But physically uprooting and lifting three-year-old plants of *O. ficus-indica* growing in California, Chile, and Israel helped convince some non-believers. Such plants, established by placing one-third of a cladode below ground level, can have 12 cladodes and weigh 15 kilograms (36 pounds) fresh weight in just three years, even when placed close together (8 plants per m^2). Visual and tactile evidence can trump statistics and numbers. Indeed, this platyopuntia can be extremely productive, even under environmentally challenging conditions.

7

Future Utilization of Agaves and Cacti

The major issues of global climate change quantified in Chapter 4 are rising atmospheric CO$_2$ levels, rising temperatures, and changing rainfall patterns. We indicated that agaves and cacti can cope with all of these eventualities (also Chapter 4). Thus only relatively minor changes based on future climates during the 21st century are expected for them on an ecological level, although any ecological change can be important and dramatic.

One ecological change expected for agaves and cacti involves increases in ranges poleward and to higher elevations. On the other hand, perhaps the most important habitat losses for agaves and cacti are in the deserts of the southwestern United States, northern Mexico, and South America. As these deserts become drier in the future, concomitant habitat losses in such regions for C$_3$ and C$_4$ plants will be even greater than for CAM plants. Also in the future, tropical regions will probably become wetter if a bit warmer, which may affect the ranges of certain epiphytic and hemiepiphytic CAM plants. But again the effects will probably be minor.

Nor are any catastrophes predicted for cultivated agaves and cacti as a result of climate change—you can still enjoy a Margarita or some nice cochineal dye. This is not to say that there will be no agricultural changes. If the locations of certain cultivated agaves and cacti are currently economically marginal for their growth, then decisions will have to be made based on principles laid out in Chapter 5 utilizing an Environmental Productivity Index. Indeed, decision making is most challenging in marginal growth regions, as is already being considered for many C_3 and C_4 crops worldwide. Witness the rush of grape growers in California and in Europe to buy lands at slightly higher elevations in anticipation of higher temperatures in the future. Also, timber companies are planting new trees taking climate change into consideration.

Although no loud alarm bells are sounding for agaves and cacti with respect to future climates, opportunities are ringing for innovative ways to use CAM plants. Indeed, CAM plants are currently underutilized worldwide in many ways that relate to climate change. *Opuntia ficus-indica* has been touted throughout this book for its great net CO_2 uptake ability (Fig. 2-4C) and its high biomass productivity (Table 6-1). But its fruits are also great and its fodder is fantastic (Chapter 1)! In this chapter we revisit the substantial biomass productivity of certain agaves and cacti, especially in regions where their high Water-Use Efficiencies based on Crassulacean Acid Metabolism (Chapter 2) are a major advantage. Actually, agaves and cacti can contribute in many ways related to dealing with the conditions and the consequences of future global climate changes.

Sinks for Global CO_2—Carbon Credits

An economic incentive for increased future uses of agaves and cacti in some regions may be the so-called *carbon credits* or the related *carbon offsets* (Capoor and Ambrosi, 2008). Beginning in Rio de Janeiro, Brazil, in 1992, and codified in Kyoto, Japan, in 1997 (the "Kyoto Protocol"), nearly all of the industrialized countries politically agreed to reduce their collective global emissions of greenhouse gases by 5.2% by the end of 2012 compared with the emissions in 1990 (it took a while for the Protocol to be ratified, such as by Canada in 2002, Russia in 2004, Australia in 2007, and so far not by the United States). This would correspond to a 29% reduction in emissions projected for 2010, if no abatement measures had been taken.

If a country cannot meet its target, which varies by country and specifically does not apply to China, India, and other developing countries, it can buy a "carbon credit" to "offset" or compensate for its excess from another country that is under its greenhouse-gas–reduction target. Or it can sponsor a greenhouse-gas reduction project in a developing country, among other options. For instance, the worldwide recession in 2007 through 2009 reduced CO_2 emissions in some regions, such as eastern Europe, as the economic downturn lessened industrial and popular use of fossil fuels. Such reductions have been sold to companies in other countries to offset their excess emission of CO_2.

Trading of carbon credits establishes a monetary value or cost for emitting greenhouse gases. This has led to the expression "cap and trade," also referred to as "emissions trading." In particular, a cap or limit is set on the emission of CO_2 or other greenhouse gases (expressed in their equivalent effect relative to CO_2) by a particular company, as determined by a governmental agency in that country. If the annual emissions fall below the set limit, the company can trade away or sell its unused carbon credits. If the company exceeds its limit, it can trade for or buy carbon credits, either privately or on a CO_2 commodity exchange.

By far the largest commodity exchange for CO_2 credits is currently the European Union Emission Trading Scheme (EU ETS). In the United States, the Chicago Climate Exchange (owned by the same company as the EU ETS, Climate Exchange PLC) is a much smaller carbon-credits trading market operating on a voluntary basis for companies, because the United States has not ratified the Kyoto Protocol. However, the U.S. House of Representatives passed legislation supporting many aspects of the Kyoto Protocol and other matters related to controlling the emission of greenhouse gases in June 2009; the U.S. Senate planned to consider this in the autumn of 2009. Trading in carbon credits is growing explosively and is predicted by many to surpass the value of all other commodities in a few years. In theory, the cap-and-trade incentive approach, where each company or government entity decides whether to reduce its CO_2 emissions or to purchase carbon credits, causes the emissions of greenhouse gases to be held in check at the least overall cost to society.

Currently, most carbon credits are sold by China and secondly by India. These are mainly to compensate for emission of hydrofluorocarbons and other industrial gases in developed countries, with relatively little influence on worldwide CO_2 emissions. In any case, a specific industry releasing much CO_2 into the atmosphere (more than its cap) can buy a carbon credit from a project or a company in a developed or a developing country that has a means of reducing its greenhouse gas emissions. Examples are switching to a more environmentally friendly energy source (carbon offset credit) or absorbing CO_2 by some means (carbon reduction credit), such as by an industrial process or by trees and other plants (more about this later).

The underlying philosophy of carbon credits is that CO_2 emission and CO_2 sequestration are reciprocal processes anywhere in the world. Therefore, emission in one place and equal sequestration in another can balance each other, leading to no effect on global climates. In this regard, agaves and cacti with their substantial biomass productivities and their high Water-Use Efficiencies should be considered for the terrestrial sequestration of atmospheric CO_2 in underexploited arid and semi-arid regions (Drennan and Nobel, 2000). Such regions, which occupy about 30% of the Earth's land area, are poorly suited to C_3 and C_4 crops without irrigation. Why not plant agaves and cacti and earn some carbon credits?

Let us next consider a way to mitigate global climate change on an individual scale. Suppose that you fly from Europe to the United States, or vice versa. Either online before departure or at check in, you can determine your carbon "footprint" or share of the CO_2 released by the aircraft's engines, which for the flight in question is usually 0.5 to 0.9 metric tonnes (1,100 to 2,000 pounds) per person, depending on the particular aircraft and the cities involved. Airlines offer various choices for carbon credits, ranging from domestic landfill utilization of methane to renewable energy solar or wind farms to planting trees in various countries. Prices currently (late 2009) range from $12 to $40 per tonne of CO_2. Some of this wide price range reflects the different costs per tonne of CO_2 saved for projects in countries such as the United States compared with opportunities in certain developing countries with lower inherent costs. Another variable affecting the price is the quality of the projects and their expected long-term sustainability, which can be difficult to evaluate.

The developed country (or you on an airplane) can thus achieve a CO_2 emission reduction strategy. At the same time, an incentive is established in developed countries to reduce greenhouse gases. Or a developing country gets capital investment or beneficial changes in land use. In any case, energy use and CO_2 emissions are continuing to increase in the 21st century for many industries, encouraging the development of environmentally more friendly endeavors, such as wind farms, that generate carbon credits to sell. Hopefully, planting agaves or cacti in Africa and other regions will be an option in the near future. Thus in addition to campaigns to plant a tree (Nair, Kumar, and Nair, 2009), we could have a CAMpaign (Osmond, 2007) to plant a CAM plant to sequester carbon.

Let's again examine this at a personal level. For instance, the annual CO_2 footprint in developed countries averages about 15 tonnes of CO_2 per person (range, 8 to 32 tonnes per person per year). Some have proposed establishing a cap-and-trade scheme on an individual basis to encourage personal conservation in developed countries. Again on a personal level, the well-being of local people in developing countries is far more important than are the needs of a company with high CO_2 emissions in a developed country that wants to obtain carbon credits. The international carbon-credit scene is changing rapidly, and many options are becoming available.

Let us apply this carbon credit concept to planting *Opuntia ficus-indica*. The value of sequestering one tonne of CO_2 from the atmosphere on the five major CO_2 commodity exchanges was about $10 in 2008, but it reached $15 per tonne in Europe in May 2009. Many would argue that its true social value then was much higher, such as $30, and in any case it will increase with inflation and increasing demand. As indicated at the end of Chapter 6, *O. ficus-indica* under modest growth conditions can have a biomass gain of 20 tonnes dry weight hectare^{-1} year^{-1} (9 tons acre^{-1} year^{-1}). One tonne of biomass, which is often mainly carbohydrates (CH_2O, Equation 2.1b), contains the same weight of carbon as about 1.5 tonnes of CO_2 (the molecular weight of CH_2O is 30 and that of CO_2 is 44, so CO_2 represents 44/30 or 1.5 times the weight of CH_2O). [To adjust to other regions or other growing conditions, the predicted productivity can be adjusted from 20 tonnes hectare^{-1} year^{-1} using EPI (Equation 5.3).]

Suppose a farmer in a region with the realistic productivity of 20 tonnes hectare^{-1} year^{-1} planted 0.1 hectare (10 meters by 100 meters, which is just over 10 yards by 100 yards) of *O. ficus-indica*. The plants could store 20 tonnes hectare^{-1} year^{-1} × 0.1 hectare or 2 tonnes of biomass (3 tonnes of CO_2 equivalent) year^{-1} for at least 20 years. The time period for carbon sequestration could be much longer, as some plantations of *O. ficus-indica* near Santiago, Chile, are 120 years old and have truly massive plants that are still growing.

Receiving $20 per tonne × 3 tonnes of CO_2 or $60 annually per 0.1 hectare for many years should be worthwhile in regions where income is low. In addition, other products such as fruits and cochineal dye could be obtained without harvesting the whole plants. Planting trees (forestation and reforestation) is already being done to obtain carbon credits. Hence, sequestering carbon in agaves and cacti probably is economically viable at the 2009 value of carbon credits and can only grow in importance in the future as the price per tonne of CO_2 increases.

Biomass for Biofuels—Cellulose Digestibility

Raising *Opuntia ficus-indica*, other opuntias, and even agaves for fodder (Chapter 1) offers probably the greatest potential for the increased land use of these plants in the future, as we will discuss later in this chapter. But what is the potential of agaves and cacti for biofuels? Ideally, a biofuel will have plants taking up CO_2 from the atmosphere and converting the CO_2 into photosynthetic products that are stored in various ways in plants. The resulting compounds are then made into a fuel such as ethanol (CH_3CH_2OH), which when burned releases CO_2 back into the air. Ignoring for the moment the CO_2 released in processing and transportation, there is then no net release of CO_2, opposite to the case of burning fossil fuels, where there is always a net release of CO_2 into the atmosphere.

Many have challenged the production of ethanol from foodstuffs that can be used to feed humans instead of to run automobiles. In particular, ethanol is produced in the United States from the kernels of the C_4 crop corn (maize, *Zea mays*) and to a small extent from the seeds of the C_3 crop soybean (*Glycine max*), both of which are heavily subsidized. In Brazil ethanol is produced mainly from sugar in the C_4 crop sugarcane (*Saccharum officinarum*). The water-saving advantages of CAM species for biofuels have so far received scant commercial attention. However, the United States Congress has mandated that at least 60% of the U.S. ethanol production is to be from crops other than corn by 2022 (Sinclair, 2009).

Environmental questions revolve around the total CO_2 costs of any crops used for ethanol. These include all of the CO_2 emitted in transportation and fertilizer usage, which can equal and hence environmentally negate the CO_2 equivalents stored as ethanol. In any case, much greater attention should be paid to converting the non-edible portions of such highly productive crops (Table 6-1) into biofuels. We could then harvest the kernels of corn for food and the cobs and stalks of corn for ethanol production. This leads to the technical problem of converting cellulose, the common structural carbohydrate polymer of plants, into monomers that can be fermented.

Cellulose is a long chain of many, many glucose subunits (often about 10,000). It represents more biomass on Earth than any other organic compound in living organisms (Nobel, 2009). Another plant polymer, lignin, is second in biomass in living organisms (about half as much as cellulose) and is even harder to break down than cellulose (coal is mainly lignin derived from plants that died long ago). Other polymers found in plants include starch, pectins, and hemicelluloses.

Breaking down or "digesting" cellulose and the other polymers on an industrial scale is done biologically using various fungi and bacteria; it is done non-biologically using acids (such as sulfuric acid) and elevated temperatures. Such difficult technology requires biochemical or chemical insight and mechanical ingenuity, and it would certainly benefit from increased research funding. A major and longstanding problem is to scale up a proven reaction in a laboratory to a commercially profitable reality (Huber and Dale, 2009).

One advantage of using the biomass from agaves and cacti to produce biofuels is that such usage can generate the carbon credits discussed in the previous section. Another and probably more important advantage is that marginal land can be utilized. Thus forests do not have to be cleared, as for sugarcane plantations in Brazil, and valuable cropland would not be usurped, as for cornfields and soybean fields in the United States. Clearing forests releases tremendous amounts of CO_2 into the atmosphere, as detailed in Chapter 4. In any case, having a productivity of 12 tonnes of biomass hectare^{-1} year^{-1} for cacti in semi-arid regions with less than 450 millimeters (18 inches) of annual rainfall (Chapter 6) is a great opportunity (see Fig. 1-1 for relatively young agaves and cacti flourishing with only 380 mm of annual rainfall). Details need to be worked out, but entrepreneurs and informed citizens will certainly see the potential.

You might ask about fermenting agave juices to produce the biofuel ethanol, thereby avoiding the problems of cellulose digestibility. However, pulque, mescal, and especially tequila are far more valuable as drinks (Chapter 1) than to pour as ethanol into your car. Moreover, the yield of such ethanol is far too low per unit ground area to be worthwhile. On the other hand, chopped cladodes of platyopuntias placed in closed containers could be used to produce methane. This has actually been done on a small scale to produce a local source of fuel for farming operations in Chile. But the real potential of agaves and cacti for biofuels rests on digesting cellulose and other polymers economically and on a large scale. This is a problem confronting the use of all types of plants for biofuels and is currently a very important and active research area.

Desertification—Our CAM Plants to the Rescue

Let us say that we can use some agaves and cacti as sinks for global atmospheric CO_2 and that somehow we eventually get a profitable way to convert their cellulose and lignin into biofuels. Are there other ways that agaves and cacti can combat the effects of global climate change? Yes. One such way is to restrain *desertification*, which refers to the degradation of multi-species ecological communities and agricultural land into less productive ones, often with few species and little ground cover. Actually, desertification has less to do with deserts per se than with degradation of soil and vegetation in dry regions, leading to impaired ecosystem structure, decreasing biodiversity, and diminished economic viability.

Desertification results in arid and semi-arid regions when over-grazing removes the local vegetation. This is the main cause of desertification in the Sahel, other parts of eastern Africa, central Asia, Australia, Mexico, and the United States (Nefzaoui and El Mourid, 2009). Desertification also results from indiscriminate plowing and excessive cultivation, which is the main cause in northwestern China, northern Africa, and the Near East. Excessive collecting of firewood has also contributed to desertification in China, other parts of Asia, and many parts of Africa.

In most cases involving desertification, humans are trying to raise too many animals on too little space. Animals range from chickens to cattle to goats, but the result is bare soil that erodes easily. Topsoil is blown away, worsening the problems. And desertification can be exacerbated by climate change, especially rising temperatures and lower rainfall in arid and semi-arid regions that already have low rainfall (Chapter 4). For instance, increasing temperatures and poor land-use practices are threatening large areas of Spain with desertification. Also, dust storms have increased in Iraq after decades of war and mis-management, compounded by drought, have increased desertification. This occurs in a region bounded by the legendary Tigris and Euphrates rivers, formerly known as Mesopotamia, the "cradle of civilization." But agaves and cacti can come to the rescue.

Planting *Opuntia ficus-indica* in the West Asia/North Africa (WANA) region, such as Algeria and Tunisia, has greatly reduced local erosion and desertification (Nefzaoui and Ben Salem, 2002). The planting of patches of *O. ficus-indica* in dry zones has been associated with small, local catchments for capturing rainfall, leading to sufficient water to sustain the plants. Also, multiple rows or hedges of *O. ficus-indica* slow topsoil and sand movement, resulting in local buildup of soil that can prevent runoff of rainfall and help restore vegetation cover. Large cactus hedges have also reduced larger-scale wind erosion, such as the massive dust storms that occur in the Sahara. Such storms blow away the lighter, more fertile topsoil, further degrading the region. The cacti can also serve as a source of food and water for animals.

Such techniques to prevent the adverse effects of habitat destruction accompanying desertification can be applied in many other parts of Africa and similar regions of the world. This is especially true where overgrazing and other practices have led to soil erosion and the loss of endemic species. In addition to being able to survive under limited nutrient and limited water resources, another advantage of planting cladodes of platyopuntias or the offshoots of agaves is their tissue water storage minimizes the problems of desiccation that lead to mortality of seedlings of shrubs and trees similarly planted under such dry conditions (De Dato et al., 2009). Preventing and reversing desertification is certainly within our grasp, and agaves and cacti can help.

Changing Climates—Regional Implications

Many of the future uses of agaves and cacti in new areas involve expansions of their utilization for fodder (Chapter 1). Fodder production already involves the most land area for all commercial CAM plants. It is crucial for raising cattle, sheep, and goats in many regions worldwide. Indeed, these animals can thrive on the cladodes of *Opuntia ficus-indica* and other platyopuntias. And such animals can feed humans.

Everyone agrees that cladodes are a great source of water in arid and semi-arid regions. But there is concern for a nutritionally balanced diet, as cladode nitrogen levels (generally slightly less than 1% of the dry weight) are usually below that for optimal growth of animals. Also, the phosphorus and the sodium contents of cladodes are too low for sustained growth without some supplements, such as bone meal that is a waste product from slaughterhouses and other industries. We will emphasize biomass productivity in various regions (see EPI, Equation 5.3), but a complete economic evaluation of local conditions requires an inclusion of nutritional requirements of the animals and the nutrient contents of the soil (Nobel, 1988; Nefzaoui and Ben Salem, 2002; Dubeux et al., 2006).

As we consider various areas for future cultivation of agaves and cacti, another question is how global climate change will increase the minimum temperatures, thereby opening up new opportunities for cultivation. For convenience, we will often refer to frost-free regions, but *O. ficus-indica* can withstand temperatures of −10°C (14°F), as discussed in Chapter 3. In any case, freezing temperatures are a real limitation to extending the cultivation of CAM plants, but breeding and biotechnology hold great promise for improving their low-temperature tolerance in the future (Chapman et al., 2002; Nobel et al., 2002; Felker et al., 2009).

Southwestern and South-central United States

The present constraints on raising agaves and cacti in the United States are partly cultural and partly climatic. By cultural constraint is meant unfamiliarity with beverages such as aguamiel and pulque, cactus fruits, and nopalitos used as a vegetable (Chapter 1). Also, use of cladodes as forage or fodder for livestock is not fully appreciated in all regions (it is common in Texas). By climatic constraint is meant the poor tolerance of these succulent plants to freezing temperatures. The former constraint is changing as people from Mexico and other parts of Latin America bring their tastes and their knowledge into the United States. The latter constraint is more relevant to the underlying theme of this book—adjusting to climate change—and will be considered next.

Currently, only about 2% of the land area of California is climatically suitable for the cultivation of *Hylocereus undatus* (mentioned in Chapter 3; Nobel et al., 2002), which is cultivated for its delicious fruits (Chapter 1). About 36% of the land area of California is currently suitable for the cultivation of *Opuntia ficus-indica* (Chapter 3; Nobel et al., 2002), which is cultivated for fodder, fruits, as the host for cochineal insects, and as a vegetable (Chapter

1). The area limitations for cultivation in California are mainly due to low wintertime temperatures.

If the temperatures were to increase by the global average of 3.1°C (5.6°F) that is predicted for the end of the 21st century (Chapter 4), then approximately 12% of the land area of California would be suitable for the cultivation of *H. undatus* and 65% for the cultivation of *O. ficus-indica*. The slightly greater temperature increases predicted for the southwestern United States compared with the global average (Christensen et al., 2007) will raise these percentages slightly. Whether advantage will be taken for the increased cultivation of these cacti in California depends on cultural and land-use practices, but the opportunity will be there.

Analogous increases in the range for agaves and cacti apply to other regions in the southwestern and south-central United States. Indeed, the southern part of the "bread basket" of the United States, where corn (maize) and wheat are currently grown to a limited extent, can become a "prickly pear" or "cactus pear" basket, as regions open up for the cultivation of platyopuntias without the threat of killer frosts. This is not so much a displacement of current crops but rather more an opportunity to raise platyopuntias on low-nutrient soils without irrigation. Thus the cultivation of agaves and cacti in the United States will inevitably increase.

Meanwhile two other changes besides *O. ficus-indica* intruding into the U.S. breadbasket may occur. First, breeding and biotechnology may develop cultivars of opuntias that are much more tolerant of freezing temperatures (Chapter 3). Second, the use of cladodes for fodder for cattle as well as for sheep, goats, and chickens, as is currently the practice in many other parts of the world, may increase in the United States. Of course, the increasing atmospheric CO_2 concentration is also good news for growing cacti in the United States in the future, as their net CO_2 uptake per unit cladode area will also be increasing (Chapter 5).

Let us next try to integrate the various factors to make predictions for the future cultivation of platyopuntias in the southwestern and the south-central United States (similar conclusions can be extrapolated to the southeastern U.S.). A 3.1°C increase in temperature will move the frost line about 350 kilometers (220 miles) north (Chapter 4). A breeding/biotechnology improvement of low-temperature tolerance of platyopuntias by a similar temperature magnitude would double this northward availability for the cultivation of these plants. Except at high elevations, growing *O. ficus-indica* would then be possible throughout most of Arizona, New Mexico, and Texas (plus much of Oklahoma, Lousiana, and Arkansas) based on the current successful cultivation of *O. ficus-indica* and *Opuntia ellisiana* in southern Texas and northern Mexico.

And what if cactus fruits and nopalitos become mainstream foods? Using nopalitos as a vegetable is five to ten times more energy efficient for human calories than is feeding the cladodes to cattle and then eating the cattle. Only time will tell how these options play out.

Latin America

A multitude of uses currently occurs for agaves and cacti in Latin America (Chapter 1). Consequently, global climate change can have many, many influences. Because most such industries are well established and are usually not in environmentally marginal areas, no major threat is perceived from the initial global warming trend or modest changes in rainfall.

For instance, the possibility of lowered rainfall in the tropics should have less effect on *Hylocereus undatus* and other pitahayas grown for their fruit than on the associated vegetation. However, increases in nighttime temperatures will tend to decrease net CO_2 uptake by most CAM plants (Fig. 5-2), especially in the warmer regions. The cactus pear orchards of Argentina and Chile in particular, and also those in Bolivia and Peru, should remain equally productive throughout the 21st century, with net CO_2 uptake benefiting from the increasing atmospheric CO_2 levels. Whether such CO_2 uptake will cause a trade-off between cladode production versus fruit production is uncertain, but most likely both organs will respond with increased growth. In any case, increases in fodder use of platyopuntias is inevitable in Latin America, based on its great success in northeastern Brazil and the climatic suitability of many other regions.

As temperatures rise, *Agave tequilana* will be able to be grown at higher elevations in Jalisco, Mexico. This will involve a simple extension of its principal cultivation region near the city of Tequila toward the north and the east. However, parts of Nayarit, Tamaulipas, and Michoacan, Mexico, where *A. tequilana* currently is grown to a limited extent, as well as south-central Jalisco, will experience higher temperatures and slightly drier situations (Christensen et al., 2007). These conditions can diminish the sugar content of the piñas somewhat, leading to less fermentable sugars that can be used to make tequila.

Actually, the greatest problems facing the monoculture of *A. tequilana* in Mexico in the future are insect pests that lay their eggs in the leaves and fungal as well as bacterial diseases that affect the leaves, the stems, and the roots. Although the physiology of this species is relatively well understood (Chapter 2), improved breeding and biotechnology efforts could make it much more ready for global climate change. In any case, purchasing or obtaining lease agreements for fields at higher elevations and hence cooler areas, as is currently being done by grape growers in California and in Europe, makes sense for growers of *A. tequilana* in Mexico. Except perhaps for *A. tequilana*, whims and changing tastes of people and the world market will probably influence the future cultivation of agaves and cacti in Latin America more than climate change.

Most of the landmass of South America lies in the tropics within 24° of the Equator. This tropical region, including the Amazon Basin, is heavily forested. Increasing temperatures and most likely decreasing rainfall will be detrimental to many tree species, leading to concern about local "dieback" (Malhi et al., 2009). But such more arid conditions are favorable for the cultivation of agaves and cacti, a topic that is relatively unexplored.

Deforestation of Amazonia has both ecological impacts and climatic implications. Deforestation leads to more exposed soil surface that reflects sunlight, leading to a local cooling effect, which is partially compensated by less plant transpiration, which will lessen cooling. Predictions of slightly decreased rainfall and slightly increased nighttime temperatures in northeastern Brazil will decrease productivity of *Opuntia ficus-indica* and *Opuntia* (*Nopalea*) *cochinellifera* growing there for fodder, which should be offset by the increase in net CO_2 uptake rate caused by the increasing atmospheric CO_2 level.

In regions of Mexico, Peru, Bolivia, and Chile where *O. ficus-indica* is grown as a host for cochineal insects (Chapter 1), precipitation may decrease slightly, with little consequence for the industry. Heavy precipitation events, especially when accompanied by high wind, can wash the insects off the cladodes in the field, but little problem is expected in the 21st century. The many influences of global climate change on agaves and cacti in Latin America need to be judged case by case.

The Sahel and Africa in General

Many ask, "Where is the Sahel?" This often poorly understood region refers to a semi-arid strip across Africa that is south of the Sahara, the largest desert in the world, and north of more temperate wooded regions. From west to east, the Sahel encompasses parts of Mauritania, Senegal, Mali, Burkina Faso, Niger, Nigeria, Chad, Sudan, Ethiopia, and Eritrea. This belt across Africa, with elevations of 200 to 400 meters (650 to 1,300 feet) and an average width of 800 kilometers (500 miles), has a long history of caravans and traders. Ecologically it is dominated by annual grasses, shrubs, and various species of *Acacia*, among other trees. Over the years, the many former indigenous species of grazing animals have largely been replaced by livestock, all of which can consume agaves or cacti as forage or fodder.

The Sahel has been subjected to severe droughts in the latter part of the 20th century. Global climate change threatens to make this region even drier in the future. Models have indicated that a major factor influencing the change in local temperatures is increases in the surface temperature of the oceans (Christensen et al., 2007). Secondarily, loss of vegetation cover (desertification) leads to higher temperatures, although the accompanying enhanced dust storms (with their aerosols) can lower temperatures somewhat. Some models predict a slight increase in rainfall in the eastern Sahel during the 21st century, but most simulations indicate slightly less precipitation overall. Because of the many countries involved and the possibility for substantial job creation, the Sahel is a keystone region for new cultivation of agaves and cacti. In this regard, *Opuntia ficus-*

indica has been successfully introduced into the eastern Sahel in Mauritania, where it can be important for combating desertification (Nefzaoui and El Mourid, 2009).

Earlier in this chapter we discussed the increasing use of *O. ficus-indica* and related species for fodder in the WANA (Western Asia/Northern Africa) region. Specifically, uses of platyopuntias for fodder have greatly expanded in Morocco, Algeria, Tunisia, Libya, and Egypt. Cactus pears are also consumed in these counties as well as in Ethiopia. These arid and semi-arid regions are excellent for such CAM plants, especially as fodder for sheep and goats. But southern Africa is also an appropriate area to raise platyopuntias for fodder. Moreover, *O. ficus-indica* has been grown in South Africa for fruit for over 100 years.

The prediction that temperatures will increase more for Africa than the global average (Christensen et al., 2007) will generally decrease net CO_2 at night by agaves and cacti (Fig. 5-2) and must be judged region by region. Of course, as the use of agaves and cacti increase in the future, so will their insect pests become more widespread (Zimmermann and Granata, 2002). Also, cochineal insects (species of *Dactylopius*) have been deliberately introduced into South Africa and Zimbabwe to control opuntias as weeds, and *Castoblastis cactorum* was also deliberately introduced into Australia to control the spread of opuntias there (discussed in the next subsection). Yet the potential of Africa for raising agaves and cacti is only beginning to be realized, and global climate change in the 21st century should have only minor direct effects on the trend. Indeed, look for many new opportunities in Africa for agaves and cacti in the future.

Australia Again

Australia has been home to some of the grandest and the most devastating ecological experiments involving exotic animals and plants. Beginning with a few introduced rabbits in the late 18th century, their population exploded until there were millions upon millions of rabbits. Also, *Opuntia stricta* and 17 other species of opuntia (especially *Opuntia ficus-indica*) came to occupy 25 million hectares (62 million acres, an area larger than the entire United Kingdom) of farmland in the east-central part of this country-continent, often with 500 tonnes of biomass dry weight per hectare (Nobel, 1994; Osmond, Neales, and Stange, 2008). This accumulation of biomass by opuntias can be construed as good news for carbon credits in semi-arid regions. The area formerly occupied by opuntias in Australia is about ten times the worldwide area currently occupied by commercial CAM plants.

The opuntias that conquered eastern Australia had humble beginnings as ornamental plants, mainly adorning vineyards, in the middle of the 19th century. They began to act as pests at the beginning of the 20th century by taking over farmland, and soon exploded with no local enemies and initially ineffective control measures. Chopping them up simply created small pieces that rooted easily. Large animals passing by detached cladodes that likewise rooted, sometimes at considerable distances away. Emus loved the fruits and rapidly spread the seeds in their feces. Over 250,000 emus were sacrificed in an effort to stop the spread of the cactus. At its peak, opuntias were spreading unchecked across Australia at 80 hectares (200 acres) per hour!

The opuntia onslaught in Australia was brought under control in the 1930s using a biological control agent, *Cactoblastis cactorum*. This moth is native to Argentina (the source of the moths brought to Australia in 1925), Paraguay, and Uruguay, whose larvae love to eat the cladodes from the inside out. But there is an incredibly valuable underlying lesson here—opuntias can grow extremely well in Australia. Indeed, this is often the case with invasive exotic species worldwide.

As indicated in Chapter 4, broadly speaking Australia can expect higher temperatures and less rainfall in the future, especially on the southern part of the continent (Christensen et al., 2007). This is not necessarily bad news for agaves and cacti. In any case, agriculture has flourished in southeastern Australia (New South Wales, Victoria, and part of South Australia), where irrigation uses most of the water from the huge Murray River and Darling River systems via a network of canals. However, drought in the early part of the 21st century, which has been exacerbated by increased temperatures, has greatly reduced the availability of such water. This is having a severe impact on conventional crops such as cotton, rice, and citrus, as contractual allotments from the government for irrigation water cannot be met. There simply is not enough water (Draper, 2009). This could portend changes in other agricultural regions of the world that have become equally heavily dependent on irrigation.

Water-conserving agaves and cacti could stage a comeback in Australia with the appropriate management practices. For *O. ficus-indica*, controlling the threat of damage by *C. cactorum* can undoubtedly be done with biological and chemical agents, but more research is necessary (Zimmermann and Granata, 2002). Perhaps this time the biomass productivity of platyopuntias can be coordinated with the appetites of cattle and sheep. Besides southeastern Australia, many other regions of Australia are frost-free, arid or semi-arid, and underutilized agriculturally. The climate of Queensland is compatible with *Hylocereus undatus* as well as other pitahayas and pitayas with delicious fruits (Chapter 1). In many ways and except for the relatively high labor costs, Australia offers great opportunities for the expansion of the cultivation of agaves and cacti in the future, taking into consideration the predictable impacts from global climate change.

Europe and Asia

Considering other regions worlwide, we bypass perennially cold Antarctica and focus on Europe and Asia. These latter two continents already have over 90% of their arable land fully occupied by human endeavors. Moreover, much of the non-arable land area in both continents is annually or occasionally submitted to temperatures below −15°C (5°F), which is inappropriate for the cultivation of agaves and cacti. Thus the "battle cry" for the future for Europe and Asia with regard to global climate change and the cultivation of agaves and cacti is to take up themes already presented in this chapter. For instance, the degradation of land by increasing desertification has already happened in both continents, ranging from much of Spain and Greece to parts of southern China. Also, as temperatures rise, future cultivation of agaves and cacti will expand into regions that are currently too cold to grow them.

Opuntia ficus-indica currently thrives throughout the Mediterranean region, with the cactus pear industry being especially well developed in Italy in general and Sicily in particular (Chapter

1). Such orchards plus more informal hedges have also been planted for a long time in semi-arid regions of Israel as well as Iran and Iraq. More recently, *O. ficus-indica* is being increasingly cultivated in Jordan and Yemen for fruits and for fodder. Syria plus regions around the Black Sea (Turkey, Bulgaria, Romania, Ukraine, Russia, and Georgia) are also suitable for the cultivation of platyopuntias (Felker et al., 2009). Global climate change will have three main climatic effects in the Mediterranean region and on into the Middle East: (1) higher temperatures, which will have major limiting effects only in the hottest regions; (2) decreases in precipitation and longer droughts, which will decrease production by agaves and cacti slightly; and (3) increased atmospheric CO_2 levels, which will increase production slightly everywhere.

For eastern and southern Asia, increases in air temperature by the end of the 21st century are predicted to be 25% less than the global average and precipitation is expected to be 8% greater (Christensen et al., 2007). Western and northern Asia will mostly remain too cold to grow agaves and cacti. However, models predict fewer extremely cold days, which is good news for these CAM plants.

In any case, niche markets for cacti in Asia will continue to prosper and most likely will expand, such as for *Hylocereus undatus* cultivated in Vietnam. This cactus has been grown in Vietnam for more than a century, is highly prized for its fruit (known as "Dragon fruit" or "Red dragon fruit"), and is extensively exported to other countries. Climate change will probably have less effect on the cultivation of this cactus than the opening of new markets in Asia for agaves and cacti. Also, wide regions of India and southern China are frost-free and suitable for the cultivation of platyopuntias for fodder, as well as for other uses (Chapter 1). Such new crops can be an important part of agricultural planning that actually can benefit from the global climates expected in the future.

Bright Ideas, Bright Future

There are so many great opportunities and so much good news for agaves and cacti in the future! Raising such Crassulacean Acid Metabolism plants on much less water than for most conventional agricultural crops (C_3 and C_4 species), especially when water availability for plants is being increasingly threatened, is just the beginning. The succulence of the leaves and the stems of CAM plants leads to additional drought tolerance. Rising temperatures will bring new frost-free areas into their possible cultivation range. Also, agaves and cacti are among the most heat-tolerant of all plants. CAM plants respond with a 1% increase in net CO_2 uptake per 10 ppm increase in atmospheric CO_2 levels; C_3 plants have a slightly lower enhancement and C_4 plants have essentially no such enhancement in net CO_2 uptake by rising atmospheric CO_2 levels (Chapter 4). Thus agaves and cacti are well positioned to cope with future global climate change.

But what about humans and their future requirements? The world population is predicted to steadily increase from 7 billion in 2010 to 9 billion in 2050 to perhaps 11 billion by the end of the 21st century (United Nations estimates). Land resources will be stressed to produce sufficient food. Platyopuntias growing in relatively infertile soils in semi-arid regions with less than 450 millimeters (18 inches) of annual rainfall can meet some of the future human food requirements with delicious fruits and a vegetable with many uses (Chapter 1).

Under restricted water conditions, which are not conducive to high productivity by conventional crops, *Opuntia ficus-indica* can produce 12 tonnes of dry matter hectare^{-1} year^{-1} (5 tons acre^{-1} year^{-1}). Its annual biomass productivity under natural conditions can be double this amount in Chile and other regions where *O. ficus-indica* is currently cultivated and can quadruple this annual biomass amount under ideal growing conditions (Chapter 6). The cladodes of this platyopuntia can be harvested as fodder for cattle, sheep, goats, and even chickens. Such planting and harvesting of cladodes of this cactus can often be accomplished with little disruption of current agricultural land uses.

Bright ideas can also be fulfilled using predictions based on an Environmental Productivity Index (EPI; Chapter 5) to assess the potential for growth of agaves and cacti in new regions. The potential for productivity can be evaluated before any offshoot of an agave or any cladode of an opuntia is placed in the ground. The establishment of these plants, which can reliably propagate vegetatively, is much more assured than planting the seeds of other plants under the hostile environmental conditions of arid and semi-arid regions.

Even regions with current cultivation of cacti for fodder can be re-assessed using EPI, as often the plants are too far apart to optimize productivity per unit ground area. Such changes in traditional practices can be done at essentially no cost. Also, for high-value crops that need some irrigation (as is often the case for *Hylocereus undatus*), an approach based upon the Water Index component of EPI can be used to evaluate the most appropriate times for water application so that the productivity per unit of applied water is maximized. Consequences of temperature changes and increasing atmospheric CO_2 levels on the productivity of agaves and cacti can also be quantified using EPI (Equation 5.3).

Another bright idea is to apply the concept of carbon credits (offsets) to growing agaves and especially to growing cacti in new regions. As indicated, *O. ficus-indica* can generate a carbon sequestration of 20 tonnes of dry matter (equivalent to 30 tonnes of CO_2) hectare^{-1} year^{-1} under realistically sub-optimal growing conditions, which can lead to substantial income in many regions worldwide. Such as in the Sahel where jobs are currently needed. This new use of agaves and cacti in arid and semi-arid areas, which can be evaluated with the information presented in previous chapters, fits well with the widely accepted goal of mitigating global climate change.

Many agaves and cacti are simply attractive. Their use as ornamentals can beautify homes and industrial establishments as well as towns and cities with little usage/wastage of water. Many of their other commercial uses are waiting to be exploited, as familiarity with their untapped potential becomes more known to the general public. Especially taking into consideration the impact of new climatic conditions expected worldwide in the 21st century, which are no great threat to them, the future of agaves and cacti is extremely bright!

GLOSSARY

absolute humidity the actual water vapor content of air, e.g., moles of water vapor per cubic meter; contrast **relative humidity**

aerosol fine solid particles or tiny liquid droplets suspended in the air, such as smoke

albedo the relative reflectivity of a land surface for sunlight

anatomy cellular and tissue structure, usually studied with a microscope

angiosperm a type of seed plant that flowers and in which the seeds form within a mature fruit; contrast **gymnosperm**

Anthropocene geologic epoch with major human influences beginning at the start of the Industrial Revolution in about 1750

AOGCM Amosphere–Ocean General Circulation Model; general model used to predict climates

apical meristem region where cell division occurs, such as at the top of the stem of a barrel or columnar cactus

areole a lateral bud on a cactus stem from which spines, glochids, and even fruits as well as stem segments are produced

arid a very dry region, usually having less than 250 millimeters (10 inches) of rainfall per year

ATP adenosine triphosphate, an important energy currency used in photosynthesis and other biochemical reactions

biochemistry chemical reactions in a plant or other living organism

biofuel energy source (often ethanol) for vehicles and factories that is derived from plants or algae

biomass plant material, usually measured in mass units after drying to remove water

bundle-sheath cells cells surrounding the vascular tissue; dark green in C_4 plants

C_3 a photosynthetic pathway whose first stable product is a three-carbon compound; used by most plants

C_4 a photosynthetic pathway whose first stable product is a four-carbon compound

CAM Crassulacean Acid Metabolism; a photosynthetic pathway used by succulent plants that has nocturnal stomatal opening and CO_2 uptake

cap and trade setting a CO_2 emissions limit and then trading (selling) carbon credits if less CO_2 is emitted (or buying if the cap is exceeded)

carbohydrate compound containing carbon, hydrogen, and oxygen; formed in photosynthesis

carbon credit/offset traded quantity, valued per tonne of CO_2 (or equivalent of another greenhouse gas), to mitigate climate change

carminic acid reddish dye produced by female cochineal insects

central vacuole a large, membrane-surrounded compartment in a plant cell

chlorenchyma green tissue of leaves or stems where photosynthesis takes place

chloroplast chlorophyll-containing organelle where photosynthesis takes place

cladode photosynthetic flattened stem segment of a platyopuntia, such as *Opuntia ficus-indica*; sometimes called a "pad"

convection movement of a liquid or gas due to local differences in density or pressure, often predominantly vertical in the atmosphere

Crassulacean Acid Metabolism CAM; a photosynthetic pathway used by succulent plants that has nocturnal stomatal opening and CO_2 uptake

cuticle waxy layer on the surface of leaves and stems that limits water loss from a plant

cuticular transpiration water loss across the cuticle of leaves and stems (the only pathway when the stomata are closed)

desert an anthropocentric term describing an extensive dry region

dicotyledon (dicot) a type of angiosperm with two cotyledons (seed leaves) on the embryo and net-like veins in the leaves or stems

dry weight biomass of plants after drying to remove water

energy currency molecule used to supply energy for a biochemical reaction, such as ATP and NADPH for photosynthesis

Environmental Productivity Index EPI; used to quantify environmental effects on net CO_2 uptake

enzyme a protein that speeds up a specific biochemical reaction

EPI *see* **Environmental Productivity Index**

epiphyte a plant growing upon another plant

family a taxonomic category consisting of related genera

fodder harvested vegetation, such as cladodes of platyopuntias, fed to animals, such as cattle, sheep, and goats

forage vegetation eaten by domestic animals in the field without harvesting

GCM General Circulation Model; used to predict present and future climates

genus (pl. **genera**) a taxonomic category containing closely related species

glacier large mass of ice on land flowing slowly by gravity from a place of accumulation to one of loss

global climate change increases in temperature, changes in rainfall amounts and timing, and changes in other climatic features worldwide

global warming observed recent increases and predicted future increases in temperature worldwide

glochid a short, thin spine with barbs that is easily dislodged from the areoles of opuntias

greenhouse gas an atmospheric gas leading to heating of the air, such as CO_2, water vapor, and methane

gymnosperm a cone-bearing seed plant lacking true flowers and having exposed ovules at the time of pollination; contrast **angiosperm**

hemiepiphyte a plant growing upon another plant but also capable of extending its roots into the soil

hydrostatic pressure force per unit area in a solution such as water that leads to flow toward lower values

inflorescence a stalk bearing multiple flowers, as for agaves

intercellular air spaces air spaces in a tissue or an organ, such as between cells in a leaf

IPCC Intergovernmental Panel on Climate Change; prepares extensive reports on future climate and on climate models

Leaf Area Index total area of the leaves (both sides for agaves) per unit ground area

mesophyll green tissue of leaves where photosynthesis takes place

methane CH_4; a potent greenhouse gas

microhabitat site occupied by a plant, etc., such as a tree crotch, whose environment can differ appreciably from that prevailing locally

monocotyledon (monocot) a type of angiosperm with one cotyledon (seed leaf) on the embryo and parallel veins in the leaves, e.g., agaves

morphology external structure and form of a plant, such as for leaves and stems

NADPH reduced form of nicotinamide adenine dinucleotide phosphate, an energy currency used in photosynthesis

NASA National Aeronautics and Space Administration; U.S. agency overseeing the space program and certain climate measurements

NOAA National Oceanic and Atmospheric Administration; U.S. agency that monitors atmospheric CO_2 levels at Mauna Loa, Hawaii

opuntia a cactus, generally with flattened stems, such as *Opuntia ficus-indica*

organelle a specialized structure, usually surrounded by a membrane, in a cell, such as a chloroplast or a mitochondrion

osmotic pressure proportional to the solute concentration in a solution; water tends to flow toward higher values (osmosis)

PEPCase phosphoenol pyruvate (PEP) carboxylase; key photosynthetic enzyme in C_4 and CAM plants
photon a particle of light, such as one absorbable by chlorophyll
photorespiration cellular process involving Rubisco that releases CO_2 in the light
photosynthesis process whereby CO_2 is fixed into a carbohydrate in chloroplasts
phyllotaxy spatial arrangement of leaves about a stem, such as the number of leaves per a certain number of complete revolutions
physiology vital functions of cells organisms, such as photosynthesis, respiration, and water movement
pitahaya edible fruit of an epiphytic or hemiepiphytic cactus
pitaya edible fruit of a columnar cactus
platyopuntia a cactus (*Opuntia*) with flattened stem segments called cladodes
PPF photosynthetic photon flux; the rate of photons incident per unit area that are absorbable by photosynthetic pigments

radiation emission and propagation of waves carrying energy, such as sunlight and longwave (thermal) = infrared radiation
RCM Regional Climate Model; used to predict the climate for a specific region
relative humidity water vapor content of air relative to its maximum or saturation content at that temperature, usually expressed in %
respiration ATP-producing process in mitochondria that requires O_2 and sugars and that releases CO_2
Rubisco ribulose-1,5-bisphosphate carboxylase/oxygenase; key photosynthetic enzyme found in all plants

semi-arid a relatively dry region, usually having 250 to 450 millimeters (10 to 18 inches) of rainfall per year
shoot aboveground portion of a plant, including the leaves (if any) and the stem or stems

"**sink**" a location to sequester something, such as atmospheric CO_2 or carbohydrates

species a taxonomic category of morphologically similar, reproductively compatible organisms derived from a single ancestral population

Stem Area Index total area of the stems per unit ground area

stomata tiny openable pores in leaves and stems through which gases such as CO_2 and water vapor pass

subtropics region between the temperate and the tropic regions

thermohaline circulation worldwide conveyor-belt-like oceanic current that includes the Gulf Stream and the North Atlantic Drift

transpiration loss of water vapor from leaves and stems

TR *see* **Transpiration Ratio**

Transpiration Ratio TR; amount of water lost by transpiration divided by the CO_2 fixation in photosynthesis; reciprocal of WUE

tropical climate continually high temperatures and considerable precipitation, as in equatorial regions

tropics region bounded by the Tropic of Cancer at 23°27'N and the Tropic of Capricorn at 23°27'S

understory the partially shaded plant region beneath the main tree canopy in a forest

Water-Use Efficiency WUE; ratio of the CO_2 fixed in photosynthesis divided by the water lost via transpiration

WUE *see* **Water-Use Efficiency**

References and Further Reading

Agaves and Cacti

Anderson, E.F. 1980. *Peyote: The Divine Cactus*. University of Arizona Press, Tucson, Arizona. 248 pp.

Anderson, E.F. 2001. *The Cactus Family*. Timber Press, Portland, Oregon. 776 pp.

Campos, F.A.P., J.C.B. Dubeux Jr., and S. de Melo Silva, eds. 2009. *Proceedings of the Sixth International Congress on Cactus Pear and Cochineal*. ISHS, Leuven, Belgium. 431 pp.

Chapman, B., C. Mondragon Jacobo, R.A. Bunch, and A.H. Patterson. 2002. Breeding and biotechnology. In: *Cacti: Biology and Uses* (P.S. Nobel, ed.). University of California Press, Berkeley and Los Angeles, California. Pp. 255–271.

Dubeux Jr., J.C.B., M.V. Ferreira dos Santos, M. de Andrade Lira, D. Cordeiro dos Santos, I. Farias, L.E. Lima, and R.L.C. Ferreira. 2006. Productivity of *Opuntia ficus-indica* (L.) Miller under different N and P fertilization in north-east Brazil. *Journal of Arid Environments* **67**: 357–372.

Felker, P., R.A. Bunch, D.M. Borchert, and J.C. Guevara. 2009. Potential global adaptivity of spineless progeny of *Opuntia ficus-indica* 1281 × O. *lindheimerii* 1250 as forage cultivars adapted to USDA cold hardiness zones 7 and 8. *Acta Horticultura* **811**: 333–342.

Garcia de Cortázar, V. and P.S. Nobel. 1990. Worldwide environmental productivity indices and yield predictions for a CAM plant, *Opuntia ficus-indica*, including effects of doubled CO_2 levels. *Agricultural and Forest Meteorology* **49**: 261–279.

Garcia de Cortázar, V. and P.S. Nobel. 1992. Biomass and fruit production for the prickly pear cactus, *Opuntia ficus-indica*.

Journal of the American Society for Horticultural Science **117**: 558–562.

Gentry, H.S. 1982. *Agaves of Continental North America*. University of Arizona Press, Tucson, Arizona. 670 pp.

Gibson, A.C. and P.S. Nobel. 1986. *The Cactus Primer*. Harvard University Press, Cambridge, Massachusetts. 286 pp.

Good-Avila, S.V., V. Souza, B.S. Gaut, and L.E. Eguiarte. 2006. Timing and rate of speciation in *Agave* (Agavaceae). *Proceedings of the National Academy of Sciences, USA* **103**: 9124–9129.

Han, H. and P. Felker. 1997. Field validation of water use efficiency of CAM plant *Opuntia ellisiana* in south Texas. *Journal of Arid Environments* **36**: 133–148.

Hartsock, T.L. and P.S. Nobel. 1976. Watering converts a CAM plant to daytime CO_2 uptake. *Nature* 262: 574–576.

Inglese, P., F. Basile, and M. Schirra. 2002. Cactus pear fruit production. In: *Cacti: Biology and Uses* (P.S. Nobel, ed.). University of California Press, Berkeley and Los Angeles, California. Pp. 163–183.

Loik, M.E. and P.S. Nobel. 1993a. Freezing tolerance and water relations of *Opuntia fragilis* from Canada and the United States. *Ecology* **74**:1722–1732.

Loik, M.E. and P.S. Nobel. 1993b. Exogeneous abscisic acid mimics cold acclimation for cacti differing in freezing tolerance. *Plant Physiology* **103**:871–876.

Metzing, D. and R. Kiesling. 2008. The study of cactus evolution: The pre-DNA era. *Haseltonia* **14**: 6–25.

Nefzaoui, A. and H. Ben Salem. 2002. Forage, fodder, and animal nutrition. In: *Cacti: Biology and Uses* (P.S. Nobel, ed.). University of California Press, Berkeley and Los Angeles, California. Pp. 199–210.

Nefzaoui, A., and M. El Mourid. 2009. Cacti: A key-stone crop for the development of marginal lands and to combat desertification. *Acta Horticultura* **811**: 365–372.

Nerd, A., N. Tel-Zur, and Y. Mizrahi. 2002. Fruits of vine and columnar cacti. In: *Cacti: Biology and Uses* (P.S. Nobel, ed.). University of California Press, Berkeley and Los Angeles, California. Pp. 185–197.

Nobel, P.S. 1980. Influences of minimum stem temperatures on ranges of cacti in southwestern United States and central Chile. *Oecologia* **47**: 10–15.

Nobel, P.S. 1988. *Environmental Biology of Agaves and Cacti.* Cambridge University Press, New York. 270 pp.

Nobel, P.S. 1989. A nutrient index quantifying productivity of agaves and cacti. *Journal of Applied Ecology* **26**: 635–645.

Nobel, P.S. 1991. Tansley Review No. 32. Achievable productivities of certain CAM plants: basis for high values compared with C_3 and C_4 plants. *New Phytologist* **119**: 183–205.

Nobel, P.S. 1994. *Remarkable Agaves and Cacti.* Oxford University Press, New York. 166 pp.

Nobel, P.S. 2001. Ecophysiology of *Opuntia ficus-indica*. In: *Cactus (Opuntia spp.) as Forage* (C. Mondragon-Jacobo and S. Perez-Gonzalez, eds.), FAO Plant Production and Protection Paper 169. FAO, Rome. Pp. 13–20.

Nobel, P.S., ed. 2002. *Cacti: Biology and Uses.* University of California Press, Berkeley and Los Angeles, California. 280 pp.

Nobel, P.S. 2003. *Agave tequilana* and tequila. *Botanical Garden UCLA* **VI**: 13–15.

Nobel, P.S., G.N. Geller, S.C. Kee, and A.D. Zimmerman. 1986. Temperatures and thermal tolerances for cacti exposed to high temperatures near the soil surface. *Plant, Cell and Environment* **9**: 279–287.

Nobel, P.S. and T.L. Hartsock. 1986. Temperature, water, and PAR influences on predicted and measured productivity of *Agave deserti* at various elevations. *Oecologia* **68**: 181–185.

Nobel, P.S., and V. Garcia de Cortázar. 1987. Interception of photosynthetically active radiation and predicted productivity for *Agave* rosettes. *Photosynthetica* **21**: 261–272.

Nobel, P.S. and A.G. Valenzuela. 1987. Environmental responses and productivity of the CAM plant, *Agave tequilana. Agricultural and Forest Meteorology* **39**: 319–334.

Nobel, P.S., E. García-Moya, and E. Quero. 1992. High annual productivity of certain agaves and cacti under cultivation. *Plant, Cell and Environment* **15**: 329–335.

Nobel, P.S., M. Castañeda, G. North, E. Pimienta-Barrios, and A. Ruiz. 1998. Temperature influences on leaf CO_2 exchange, cell viability and cultivation range for *Agave tequilana*. *Journal of Arid Environments* **39**: 1–9.

Nobel, P.S. and E. De la Barrera. 2002a. High temperatures and net CO_2 uptake, growth, and stem damage for the hemiepiphytic cactus *Hylocereus undatus*. *Biotropica* **34**: 225–231.

Nobel, P.S. and E. De la Barrera. 2002b. Stem water relations and net CO_2 uptake for a hemiepiphytic cactus during short-term drought. *Environmental and Experimental Botany* **48**: 129–137.

Nobel, P.S., E. De la Barrera, D.W. Beilman, J.H. Doherty, and B.R. Zutta. 2002. Temperature limitations for cultivation of edible cacti in California. *Madroño* **49**: 228–236.

Nobel, P.S. and E. De la Barrera. 2004. CO_2 uptake by the hemiepiphytic cactus, *Hylocereus undatus*. *Annals of Applied Biology* **144**: 1–8.

Nobel, P.S. and B.R. Zutta. 2008. Temperature tolerances for stems and roots of two cultivated cacti, *Nopalea cochenillifera* and *Opuntia robusta*: Acclimation, light, and drought. *Journal of Arid Environments* **72**: 633–642.

Pimienta-Barrios, Eu., J. Zañudo, E. Yepez, En. Pimienta-Barrios, and P.S. Nobel. 2000. Seasonal variations of CO_2 uptake for cactus pear (*Opuntia ficus-indica*) and pitayo (*Stenocereus queretaroensis*) in a semi-arid environment. *Journal of Arid Environments* **44**: 73–83.

Pimienta-Barrios, E., C. Robles-Murguia, and P.S. Nobel. 2001. Net CO_2 uptake for *Agave tequilana* in a warm and a temperate environment. *Biotropica* **33**: 312–318.

Raveh, E., M. Gersani, and P.S. Nobel. 1995. CO_2 uptake and fluorescence responses for a shade-tolerant cactus *Hylocereus undatus* under current and doubled CO_2 concentrations. *Physiologia Plantarum* **93**: 505–511.

Sáenz-Hernández, C., J. Corrales-García, and G. Aquino-Pérez. 2002. Nopalitos, mucilage, fiber, and cochineal. In: *Cacti: Biology and Uses* (P.S. Nobel, ed.). University of California Press, Berkeley and Los Angeles, California. Pp. 211–234.

Ting, I.P., and S.R. Szarek. 1975. Drought adaptation in crassulacean acid metabolism plants. In: *Environmental Physiology of Desert Organisms* (N.F. Hadley, ed.). Dowden, Hutchinson & Ross, Stroudsburg, Pennsylvania. Pp. 152–167.

Valenzuela-Zapata, A.G. and G.P. Nabhan. 2004. *Tequila: A Natural and Cultural History.* University of Arizona Press, Tucson, Arizona. 113 pp.

Wallace, R.S. and A.C. Gibson. 2002. Evolution and Systematics. In: *Cacti: Biology and Uses* (P.S. Nobel, ed.). University of California Press, Berkeley and Los Angeles, California. Pp. 1–21.

Zimmermann, H.G., and G. Granata. 2002. Insect pests and diseases. In: *Cacti: Biology and Uses* (P.S. Nobel, ed.). University of California Press, Berkeley and Los Angeles, California. Pp. 235–254.

CAM, WUE, and EPI

Black, C.C. and C.B. Osmond. 2003. Crassulacean acid metabolism photosynthesis: 'working the night shift.' *Photosynthesis Research* **76**: 329–341.

Drennan, P.M. and P.S. Nobel. 2000. Responses of CAM species to increasing atmospheric CO_2 concentrations. *Plant, Cell and Environment* **23**: 767–781.

Nobel, P.S. 2009. *Physicochemical and Environmental Plant Physiology*, 4th ed. Academic Press/Elsevier (Copyright Elsevier), San Diego, California. 582 pp.

Osmond, C.B. 2007. Crassulacean acid metabolism: Then and now. *Progress in Botany* **68**: 1–32.

Osmond, B., T. Neales, and G. Stange. 2008. Curiosity and context revisited: crassulacean acid metabolism in the Anthropocene. *Journal of Experimental Botany* **59**: 1489–1502.

Raven, J.A. and R.A. Spicer. 1996. The evolution of Crassulacean acid metabolism. In: *Crassulacean Acid Metabolism: Biochemistry, Ecophysiology and Evolution* (K. Winter and J.A.C. Smith, eds.). Springer–Verlag, Berlin. Pp. 360–385.

Winter, K. and J.A.C. Smith, eds. 1996. *Crassulacean Acid Metabolism: Biochemistry, Ecophysiology and Evolution.* Springer–Verlag, Berlin. 449 pp.

Climate/Climate Change

Capoor, K., and P. Ambrosi. 2008. *State and Trends of the Carbon Market 2008.* World Bank Institute, Washington, D.C. 78 pp.

Chapin III, F.S., P.A. Matson, and H.A. Mooney. 2002. *Principles of Terrestrial Ecosystem Ecology.* Springer, New York. 436 pp.

Christensen, J.H., B. Hewitson, A. Busuioc, A. Chen, X. Gao, I. Held, R. Jones, R.K. Kolli, W.-T. Kwon, R. Laprise, V. Magaña Rueda, L. Mearns, C.G. Menéndez, J. Räisänen, A. Rinke, A. Sarr, and P. Whetton. 2007. Regional Climate Predictions. In: *Climate Change 2007: The Physical Basis. Contribution of Working Group I to the Fourth Assessment Report of the Intergovernmental Panel on Climate Change* (S. Solomon, D. Qin, M. Manning, Z. Chen, M. Marquis, K.B. Averyt, M. Tigmor, and H.L. Miller, eds.). Cambridge University Press, Cambridge, UK, and New York. Pp. 847–940.

De Dado, G.D., L. Loperfido, P. De Angelis, and R. Valentini. 2009. Establishment of a planted field with Mediterranean shrubs in Sardinia and its evaluation for climate mitigation and to combat desertification in semi-arid regions. *iForest* **2**: 77– 84.

Draper, R. 2009. Australia's dry run. *National Geographic* (April): 34–59.

Huber, G.W. and B.E. Dale. 2009. Grassoline at the pump. *Scientific American* (July): 52–59.

Kolbert, E. 2009. Outlook: Extreme. *National Geographic* (April): 60–65.

Malhi, V., L.E.O.C. Aragão, D. Galbraith, C. Huntingford, R. Fisher, P. Zelazowski, S. Stich, C. McSweeney, amd P. Meir. 2009. Exploring the likelihood and mechanism of a climate-change-induced dieback of the Amazon rainforest. *Proceedings of the National Academy of Sciences, USA,* in press.

Nair, P.K.R., B.M. Kumar, and V.D. Nair. 2009. Agroforestry as a strategy for carbon sequestration. *Journal of Plant Nutrition and Soil Science* **172**: 10–23.

Rahmstorf, S. 2006. Thermohaline ocean circulation. In: *Encyclopedia of Quaternary Sciences* (S.A. Elias, ed.). Elsevier, Amsterdam. Pp. 1–10.

Randall, D.A., R.A. Wood, S. Bony, R. Colman, T. Fichefet, J. Fyfe, V. Kattsov, A. Pitman, J. Shukla, J. Srinivasan, R.J. Stouffer, A. Sumi, and K.E. Taylor. 2007. Climate models and their evaluation. In: *Climate Change 2007: The Physical Basis. Contribution of Working Group I to the Fourth Assessment Report of the Intergovernmental Panel on Climate Change* (S. Solomon, D. Qin, M. Manning, Z. Chen, M. Marquis, K.B. Averyt, M. Tigmor, and H.L. Miller, eds.). Cambridge University Press, Cambridge, U.K., and New York. Pp. 589–662.

Sinclair, T.R. 2009. Taking measure of biofuel limits. *American Scientist* **97**: 400–407.

Smith, N.V., S.S. Saatchi, and J.T. Randerson. 2004. Trends in high northern latitude soil freeze and thaw cycles from 1988 to 2002. *Journal of Geophysical Research* **109**: D12,101–D12,114.

www.laalamanac.com/weather: Los Angeles rainfall records

www.mlo.noaa.gov: Atmospheric CO_2 levels

www.wikipedia.org: Carbon credit; Carbon offset; Chicago Climate Exchange; Climate change; Emissions trading; General Circulation Models; Glacier; Glacier National Park; Global warming; Greenhouse effects; Gulf Stream; International Panel on Climate Change; IPCC Fourth Assessment Report; Kyoto Protocol; Mauna Loa Observatory; Methane; NASA; Retreat of glaciers since 1850; Temperature record; Thermohaline circulation; Volcano

www.earthobservatory.nasa.gov: Thermohaline circulation

Other Topics

Epstein, E., and A.J. Bloom. 2005. *Mineral Nutrition of Plants: Principles and Perspectives*, 2nd ed. Sinauer, Sunderland, MA. 400 pp.

Nobel, P.S. 1989. Shoot temperatures and thermal tolerances for succulent species of *Haworthia* and *Lithops*. *Plant, Cell and Environment* **12**: 643–651.

www.wikipedia.org: Carboniferous; Compari; Cretaceous; Diabetes mellitus; Fibonacci; Nylon; Prohibition; Sahel; Synthetic fiber

Index

C

E

east–west orientation, cladode 104
Echinocactus grusonii 26
Ecuador 16, 25
Egypt 156
elevation 72, 73, 112, 113, 115, 142, 153, 154
El Salvador 132
energy currencies 133, 134, 165
Environmental Productivity Index 99, 100, 107, 111, 114, 121, 125, 142, 160, 165
enzyme 32, 36, 134, 135, 165, 167
epiphytes 15, 31, 51, 52, 57, 165
Epithelantha bokei 66
ethanol 6, 7, 147, 148, 163
Ethiopia 24, 155, 156

F

fern 52, 53
Ferocactus acanthodes 57, 61, 63, 64, 72, 97, 103, 106, 109, 116
fertilizer 26, 117, 118, 147
Fibonacci Series 105, 111
fire 80, 131
fluorescent lamps 29, 101
fodder 1, 2, 3, 10, 19, 20, 27, 75, 136, 139, 146, 150, 151, 152, 153, 154, 155, 156, 159, 160, 161, 165
forage 1, 19, 20, 75, 155, 165
fossil fuels 81, 85, 143, 147
freezing temperatures 7, 12, 15, 16, 20, 67, 68, 70, 71, 73, 74, 96, 97, 151, 152
fresh weight 124, 140
frost 86, 96, 139, 152, 159, 160
fructose 8, 17

G

gas exchange measurement 37, 39, 40

General Circulation Model 85, 86, 90, 165
glaciers 86, 89, 90, 91
global climate change 25, 26, 36, 50, 52, 55, 60, 74, 85, 87, 90, 91, 93, 96, 99, 131, 141, 142, 144, 149, 151, 153, 154, 155, 156, 158, 160, 161, 165
global warming 81, 84, 91, 135, 153, 165
glochids 14, 16, 17, 75, 116, 163, 165
Glycine max 147
goats 10, 11, 18, 20, 149, 150, 152, 156, 160, 165
Greece 158
greenhouse gas 83, 84, 86, 87, 89, 143, 144, 164, 165, 166
Guatemala 16
Guinness World Records 60, 67
Gulf Stream 92, 93, 168

H

hemiepiphytes 15, 31, 51, 52, 57, 166
high temperature 27, 61, 62, 64, 65, 66, 67, 76, 96, 135
hormones 11, 25, 71
horses 20
hydrostatic pressure 56, 57, 166
Hylocereus undatus 15, 16, 65, 67, 74, 75, 103, 104, 106, 109, 151, 153, 158, 159, 161

I

India 9, 29, 90, 143, 144, 159
Indonesia 81
Industrial Revolution 79, 80, 93, 163
inflorescence 3, 4, 7, 10, 12, 43, 166
intercellular air spaces 47, 48, 49, 70, 71, 166
IPCC 85, 86, 89, 166
Iran 159

O

oceans 80, 82, 87, 91, 93, 155
Opuntia amyclea 132
Opuntia basilaris 60
Opuntia bigelovii 19
Opuntia cylindrica 25
Opuntia ellisiana 139, 153
Opuntia engelmannii 18
Opuntia ficus-indica 13, 15, 16, 18, 21, 22, 41, 43, 44, 47, 53, 64, 67, 73, 74, 95, 97, 103, 106, 109, 117, 125, 127, 131, 132, 133, 136, 142, 145, 146, 149, 150, 151, 154, 156, 158, 160, 164, 166
Opuntia fragilis 67, 70, 71, 74
Opuntia robusta 16, 19
Opuntia stricta 156
Orchidaceae 52
osmotic pressure 57, 68, 167
oxygen 4, 28, 36, 80, 134, 164
ozone 84, 88

P

Paraguay 157
PEPCase 32, 33, 34, 46, 47, 167
Pereskia 52
Peru 22, 24, 25, 153, 155
peyote 24
pH 21, 34, 118
Philip L. Boyd Deep Canyon Desert Research Center 112, 137
phosphorus 20, 117, 118, 137, 150
photons 102, 120, 167
photorespiration 36, 37, 51, 79, 80, 82, 94, 134, 135, 167
photosynthesis 4, 28, 29, 32, 34, 35, 36, 37, 41, 42, 46, 47, 50, 51, 58, 79, 80, 82, 93, 94, 101, 102, 104, 108, 123, 125, 126, 130, 133, 134, 135, 163, 164, 165, 166, 167, 168
photosynthetic photon flux 102, 103, 167

photosynthetic productivity 28, 80
phyllotaxy 111, 167
pigs 2, 20
piña 7, 8, 43, 131, 154
pineapple 7, 35, 133
pitahayas 15, 16, 153, 158, 167
pitayas 15, 158, 167
plant physiological ecology 39
platyopuntias 2, 16, 19, 20, 24, 150, 167
Portugal 132
potassium 117, 118, 137
prohibition 5
pubescence 71, 72
pulque 3, 4, 5, 116, 148, 151
pulqueria 5

R

rainfall 2, 7, 15, 20, 27, 56, 57, 60, 88, 89, 90, 95, 97, 108, 110, 113, 116, 136, 137, 138, 141, 148, 149, 150, 153, 154, 155, 157, 160, 163, 165, 167
Rebutia 26
Red Dragon fruits 16
reforestation 146
relative humidity 39, 40, 47, 48, 89, 163, 167
Rhipsalis baccifera 31
rice 75, 84, 97, 157
roots 15, 56, 57, 59, 95, 108, 110, 124, 125, 154, 166
Rubisco 31, 32, 33, 34, 36, 41, 46, 47, 94, 95, 134, 135, 167
Russia 143, 159

S

Saccharum officinarum 30, 132, 147
saguaro 26, 61
Sahara 87, 90, 139, 150, 155
Sahel 149, 155, 161
saline soil 118
San Pedro cactus 25
sapogenins 11

seedling 53, 97, 150
Selenicereus megalanthus 15
semi-arid regions 15, 31, 116, 148,
    157, 159, 160, 167
Senegal 20, 155
sheep 11, 18, 20, 137, 150, 152, 156,
    158, 160, 165
Sicily 13, 15, 158
sisal 9
sodium 20, 118, 150
Sonoran Desert 19, 44, 48, 57, 60, 61,
    72, 97, 108, 112, 113, 114, 137
sorghum 30, 33, 41
Sorghum bicolor 30, 132
South Africa 13, 24, 30, 61, 66, 75,
    132, 156
soybean 75, 147, 148
Spain 22, 149, 158
spineless 19, 75
spines 2, 8, 12, 13, 16, 17, 19, 26, 71,
    72, 116, 163
Sri Lanka 31
Stem Area Index 127, 128, 129, 130,
    139, 168
Stenocereus queretaroensis 15
Stenocereus thurberi 61
stomata 33, 34, 35, 37, 39, 41, 42, 43,
    45, 46, 49, 50, 51, 59, 94, 105,
    164, 168
succulence 108, 109, 160
sugarcane 30, 33, 147, 148
sunlight 16, 85, 87, 89, 101, 104, 123,
    125, 126, 128, 130, 131, 137,
    154, 163, 167
supercool 69, 70

T

temperature 40, 42, 48, 86, 105
Temperature Index 101, 105, 106,
    107, 121, 129, 138, 139
tequila 5, 6, 7, 43, 75, 116, 148, 154
Texas 18, 19, 24, 66, 107, 139, 151,
    153

Thailand 16
thermohaline circulation 91, 93, 168
Tillandsia 31
transpiration 34, 37, 38, 42, 43, 44,
    45, 46, 47, 48, 49, 50, 51, 58,
    59, 94, 154, 168
Transpiration Ratio 46, 47, 50, 168
Trichocereus chiloensis 72
Trichocereus pachanoi 25
tungsten lamps 29, 101
Tunisia 13, 20, 136, 149, 156

U

United States 5, 15, 30, 73, 75, 80,
    88, 90, 132, 141, 143, 147, 148,
    149, 151, 152
Uruguay 157
Utah 68

V

Vietnam 15, 16, 107, 159
volcanic eruptions 84, 85, 88

W

Water Index 101, 107, 108, 109, 110,
    113, 114, 116, 121, 129, 138,
    139, 161
Water-Use Efficiency 42, 46, 49, 50,
    51, 59, 95, 106, 120, 125, 136,
    168
water vapor content, saturated 48
wheat 41, 75, 97, 152

Y

Yemen 159
Yucatán 9, 115, 116, 125

Z

Zea mays 30, 132, 147
Zimbabwe 30, 156